Responsible Conduct with Animals in Research

RESPONSIBLE CONDUCT with ANIMALS in RESEARCH

edited by
LYNETTE A. HART

New York Oxford
Oxford University Press
1998

Oxford University Press

Oxford New York
Athens Auckland Bangkok Bogotá Buenos Aires Calcutta
Cape Town Chennai Dar es Salaam Delhi Florence Hong Kong Istanbul
Karachi Kuala Lumpur Madrid Melbourne Mexico City Mumbai
Nairobi Paris São Paulo Singapore Taipei Tokyo Toronto Warsaw

and associated companies in
Berlin Ibadan

Published by Oxford University Press, Inc.
198 Madison Avenue, New York, New York 10016

Oxford is a registered trademark of Oxford University Press

Library of Congress Cataloging-in-Publication Data
Responsible conduct with animals in research / edited by Lynette A. Hart.
p. cm.
Includes bibliographical references (p.) and index.
ISBN 0-19-510511-7; ISBN 0-19-510512-5 (pbk.)
1. Animal experimentation—Moral and ethical aspects. 2. Animal
welfare—Moral and ethical aspects. 3. Laboratory animals.
I. Hart, Lynette A.
HV4915.R47 1998
179'.4—dc21 97-35388

9 8 7 6 5 4 3 2 1

Printed in the United States of America
on acid-free paper

Dedicated to all the animals,
especially laboratory mice and rats

Contents

Contributors

Arnold Arluke, Professor, Department of Sociology and Anthropology, Northeastern University, Boston

Kathryn A. Bayne, Associate Director, Association for Assessment and Accreditation of Laboratory Animal Care International, Rockville, Maryland

Marc Bekoff, Professor, Department of Environmental, Population, and Organismic Biology, University of Colorado, Boulder

Gordon M. Burghardt, Professor, Department of Psychology, University of Tennessee, Knoxville

Marian Stamp Dawkins, University Research Lecturer, Animal Behavior Research Group, Department of Zoology, University of Oxford

Donald A. Dewsbury, Professor, Department of Psychology, University of Florida, Gainesville

John P. Gluck, Professor, Department of Psychology, University of New Mexico, Albuquerque

Julian Groves, Professor, Division of Social Science, The Hong Kong University of Science and Technology

Lynette A. Hart, Director, Center for Animals in Society and UC Center for Animal Alternatives, School of Veterinary Medicine; and Associate Professor, Population, Health, and Reproduction, University of California, Davis

Harold A. Herzog, Professor, Department of Psychology, Western Carolina University, Cullowhee

Melinda A. Novak, Professor, Department of Psychology, University of Massachusetts, Amherst

Andrew N. Rowan, Senior Vice President, Humane Society of the United States, and Adjunct Professor, Tufts School of Veterinary Medicine

Stephen J. Suomi, Director, Laboratory of Comparative Ethology, National Institute for Child Health and Human Development, Bethesda, Maryland

John G. Vandenbergh, Professor, Department of Zoology, North Carolina State University, Raleigh, North Carolina

Meredith J. West, Professor, Departments of Psychology and Biology, Indiana University, Bloomington

LYNETTE HART

Introduction

To address misrepresentation and outright fraud in science, along with fabrication and plagiarism of data, the scientific community has been educating both its workers and the public about their responsibilities. New books address the responsible conduct of science and even include case studies and scenarios likely to apply to animal behaviorists. However, to date no major publications have specifically addressed the particular dilemmas and issues of relevance to the field of animal behavior. To address this need, this book has evolved from an ongoing lecture series on the responsible conduct of science at the University of California, Davis, cosponsored by the NSF Research Training Grant in Animal Behavior and the University of California Center for Animal Alternatives.

Issues for scientists who work in biotechnology and animal behavior sometimes differ. For example, a common dilemma for those working in molecular biology and pharmacology involves maintaining ethical responsibilities to both academic employers and industrial firms with which they may consult. The financial stakes sometimes associated with a conflict of interest in biotechnology generally do not exist in animal behavior, where the issues are different. The field researcher, for example, after several years of data collection, may feel confronted with the ethics of certain practices involving ownership of data and assignment of authorship for projects, especially when many investigators have been involved.

Not surprisingly, those working in animal behavior have concerns for the well-being of the animals they study. They are sensitized to the possible suffering of animals from research practices in the laboratory or captive setting. In the field, the desire to gain information that will ultimately play a role in sustaining wild populations might have to be weighed against the cost of inflicting some level of discomfort on individuals in the capture or selected shooting for taking biopsies to obtain the necessary information. Because laypeople in society often identify with the emotional reactions of animals, these problems are escalated by social controversy, which is ever present for those working in the field of animal behavior.

This edited book is presented in four parts. Part I concerns changing research practices regarding animal care and use from both institutional and personal perspectives. Many research practices involving animals that were common 50 years ago may now be considered only marginally acceptable. Within the United States, the ever changing Animal Welfare Act represents a legislative response on behalf of society to the widespread social changes in attitudes to animals. Three chapters in this section provide examples of the changing patterns of acceptable animal use in law and within the American Psychological Association and of the changing perspective with which scientists may view some of their past work. Lynette Hart outlines the evolution of regulatory frameworks concerning animal use in the United States and abroad. Don Dewsbury documents the development of the animal care committee within the American Psychological Association. Among the laboratory researchers who have engendered negative public response regarding animal use, Harry Harlow, for his work with rhesus monkeys, stands out as one of the most notorious. John Gluck looks back on his experiences within Harry Harlow's laboratory and offers an enlightening current perspective.

Part II includes four chapters that focus on current concerns about the responsible conduct of research and how scientists make practical decisions pertaining to the animals they work with. Several well-publicized examples of scientific misconduct in the past few years have stimulated a strong emphasis on the responsible conduct of science. Although plagiarism, fraud, and fabrication are types of misconduct that can arise in any scientific discipline, rules about animal well-being and approaches to addressing well-being are prominent only in the field of animal behavior. These and other issues that are relevant to the field of animal behavior are addressed in this section. Melinda Novak and colleagues present a general framework for the ethical conduct of science in the field of animal behavior. Their chapter includes a list of guidelines for the use of animals that have been adopted by various professional organizations. John Vandenbergh addresses the institutional concerns for animal welfare that are overseen by the institutional animal care and use committees. The final two chapters bridge the emerging basic research area of animal cognition and animal awareness with the issues of animal use. Clearly, ethologists should play a major role in addressing this area and developing guidelines that enhance animal well-being. Both Gordon Burghardt and Marc Bekoff argue for ethologists to take a broader perspective that includes drawing from an informed anthropomorphism or even folk psychology when making decisions regarding animals.

Part III addresses the possibility of assessing an animal's well-being, preferably by asking the animal. The central issue in animal use concerns measuring or evaluating the degree of well-being of the animal in the laboratory and field, in captive or home environments. The two chapters in this section concern the identification of principles for assessing an animal's well-being. Andrew Rowan argues that animal suffering is the key issue and critically discusses differences between suffering, pain, and anxiety. Marian Dawkins recommends that the animal is the ultimate authority on its own welfare and points out methods for allowing the animal to answer questions about its own welfare.

Part IV addresses the influences that research animals can have on humans. The two chapters in this section focus on specific groups of people who are affected by their involvement with animals: animal researchers, animal caregivers, and animal protection activists. The chapters focus on several interesting topics: (1) that animals can affect the humans who are doing research with them, although those affected may not be aware of the animals' impact on them; (2) that animal welfare has become a major social issue; and (3) that politicizing animal care affects everyone who has a relationship with animals. Arnold Arluke and Julian Groves document the stress experienced by animal care personnel, particularly that associated with euthanasia of animals. Hal Herzog describes the characteristics of a group of animal rights activists ranging from moderates to extremists.

Each of the contributions to this book has been subjected to careful peer review prior to publication. In this regard, I wish to thank Katherine Bayne, Marc Bekoff, Ned Buyukmihci, Dick Coss, Don Dewsbury, Hal Herzog, Dale Lott, Bill Mason, Joy Mench, Robert Murphey, Don Owings, Andrew Rowan, James Serpell, Ken Shapiro, Robert Sommer, Phil Tillman, and Steve Zawistowski. Judith Grumstrup-Scott willingly coordinated preparation of the final manuscript. David Anderson contributed relevant bibliographic information. I especially want to thank Kirk Jensen, senior editor at Oxford University Press, and all of the contributors for their patience and assistance in completing this project.

PART I

CHANGING RESEARCH PRACTICES REGARDING ANIMAL CARE AND USE

Institutional and Personal Perspective

1

LYNETTE A. HART

Responsible Animal Care and Use

Moving Toward a Less-Troubled Middle Ground

Initiatives to improve the care of laboratory animals and encourage and develop alternatives to procedures that cause discomfort to animals are accelerating around the world (Mayer, Whalen, and Rheins, 1994). Over time, informal monitoring systems concerning animal use within various countries have become formalized as regulatory processes. The legislative and regulatory systems of the various countries contrast strongly with regard to centralized or local oversight, the requirement for licensure, whether legislated or managed through assigned professional or institutional responsibilities, ethical issues, the species covered, and the extent of inventoried record keeping. Although the systems of various countries differ in design, their goals are similar and provide evidence of an evolving and growing concern for animal well-being. In recent years, the systems within countries have been revised and updated to accommodate the evolving perceptions and expectations of what is acceptable animal care and to bring countries into conformity with each other, a process termed *harmonization.* As a primary example, the European Community (EC) stands out in its efforts to effect harmonization and set mutual goals among countries for their regulatory practices.

Robert Garner, in describing the political science of animal research, recently characterized it as the most important moral issue of our time (Garner, 1996). Within recent history in the United States, according to Garner, the animal protection movement has grown, public hostility has arisen and increased as a result of several critical incidents involving animals, and a countermobilization has occurred within the research community. His analysis emphasizes the alienation of the animal protection movement in the United States from the scientific community. Slower than some European countries to begin informed discussion, develop consensus, and pass relevant legislation, and being a highly pluralistic society, the United States today lags in some of its regulatory practices regarding animals. Garner concludes that decision making in the United States is less based on ethics than in some other countries. There is less public will to enact legislation. Many Americans dislike increasing regulation, and government agencies are influential and allied with the research community in limiting implementation. In contrast, the United Kingdom's mandatory licensing, centralized enforcement, and cost-benefit assessment that requires ethical judgment have combined with a search for compromise among the stakeholders and a responsiveness for reform.

This chapter evolves from the U.S. context, where the largest amount of research involving animals occurs. The tremendous complexity and magnitude of animal use in the United States perhaps account for so many people who feel they belong to "the troubled middle" with regard to animal use, rather than identifying with a clear-cut ethical position (Donnelley, 1990). With the objectives of presenting constructive options that may be useful and profiling some contrasts among countries, this chapter outlines some strategies countries are using to address the troubled middle ground. The chapter is not intended to comprehensively review the subject, but rather to highlight some actions being used to improve animal care and use practices.

Academic Leadership in Animal Behavior and Welfare

Ethology, or the study of animal behavior, is the basic academic discipline for assessing animal well-being. Beginning with Darwin, who documented the behavior and emotions of animals (Darwin, 1872/1965), and continuing with the Nobel Prize–winning work of Tinbergen (1972) on black-headed gulls and Lorenz (1991) in characterizing the individuality of the personalities and life stories of various greylag geese, classic studies have put forward the principle that the full range of an animal's behavior is important and that restricting its behavior is a kind of deprivation. Traditionally, the field of animal behavior studies animals in their natural context with minimal intervention whenever possible; this kind of work provides essential baseline information for understanding what is essential for the animal's well-being.

United Kingdom

The Universities Federation of Animal Welfare (UFAW) in England, founded in 1926, has long provided organizational support within universities for constructive discussion of animal welfare. Among other publications, UFAW has published the journal *Animal Welfare* beginning in 1992. Additionally, the Fund for the Replacement of Animals in Medical Experiments (FRAME) was founded in 1969 and has published *Alternatives to Laboratory Animals* (*ATLA*) since 1973. Because FRAME's aim of eliminating the need for live animal experiments depends on the proper development and use of replacement alternative methods, FRAME is actively involved in research. Both of these organizations have worked closely with the academic community to further animal welfare.

Over the years, scholarly leadership emerged at two of England's most eminent universities. From Oxford University, Marian Dawkins in *Animal Suffering* (1980) made the argument that scientists could ask animals direct questions concerning their welfare; she specified sample methods for the scientific investigation of animal welfare. Cambridge University took a bold step at its veterinary school with the establishment of an endowed chair in animal welfare and the appointment of Donald Broom to the position in 1986, suggesting that studies of animal welfare would continue to be important for the foreseeable future. (Indeed, veterinary schools in the United Kingdom now typically have a faculty position in animal welfare.) Both Dawkins and Broom have accepted graduate students who have conducted studies and contributed to the growth of knowledge of animal welfare. They serve as prominent national spokespersons from academe who speak out for animal welfare and contribute to discussions of legislation.

The English cultural tradition of teatime is a scheduled daily routine that in universities is used for intellectual conversation on any topic of interest. This practice provides a time for thinking in new areas and talking with other departmental faculty. This tradition fosters the art of discussion of varying viewpoints and permits consideration of concepts of animal welfare or other controversial topics.

United States

Although the United States has some strong academic spokespersons for animal welfare, the field has not developed the stature and leadership equivalent in impact to that of Dawkins and Broom in the smaller U.K. environment. However, the Animal Behavior Society (ABS), the professional society for scientists in animal behavior and copublisher of *Animal Behaviour* with its sister organization in the United Kingdom, the Association for the Study of Animal Behaviour, actively considers policy issues on animal welfare. Since 1977, an ABS Animal Care Committee has participated several times in refining "Guidelines for the Treatment of Animals in Behavioural Research and Teaching" (1994) to reflect the increasing sensitivity to what are appropriate interventions with animals. Methodologies with animals in submitted papers are screened for appropriateness, and questionable practices are reviewed by the Animal Care Committee.

Scientists who conduct behavioral and ecological field studies have become increasingly conscious of their own impacts on animals, prompting ethical reevaluation of various practices in collecting animals (Farnsworth and Rosovsky, 1993). When one field biologist experimentally induced infanticide by jacanas (Emlen, 1993), the work reported in the published paper was strongly criticized (Bekoff, 1993). This public examination of field research methods stimulated conversations within the field of animal behavior and contributed to the ongoing examination and revision of guidelines. The most recent revision to the "Guidelines for the Treatment of Animals in Behavioural Research and Teaching" (1994) specifically singles out for scrutiny procedures that result in, for example, aggression, predation, isolation, and crowding.

One active movement in the U.S. scientific community that could have included animal welfare in its purview is termed *the responsible conduct of science*. Rather, it has focused almost solely on problems in data management, including fraud, plagiarism, and fabrication (Schachman, 1993), and conflicts of interest for scientists, especially with regard to financial issues. *Science* has regularly reported episodes on such classic cases of possible fraud or plagiarism as those involving Robert Gallo and David Baltimore. To address these problems, the Office of Scientific Integrity was established by the U.S. Department of Health and Human Services in 1989 and succeeded by the Office of Research Integrity. New official rules for financial disclosure to avoid a conflict of interest (Anderson, 1994) and updated definitions of misconduct (Kaiser, 1995) have been published. Some faculty have developed curricular materials for teaching this material to graduate and postdoctoral students, for example, at Harvard Medical School, Georgetown University, and Indiana University. The National Institutes of Health (NIH) and the National Science Foundation have encouraged and sometimes required doctoral and postdoctoral students to take courses on the responsible conduct of science, especially when the institution has received a training grant. Although these activities help to raise ethical sensitivities, discussions on the responsible conduct of science have remained focused on individual scientists' misbehaviors regarding data and funding. They have not included laboratory animal welfare as an issue. The absence of animal welfare as a topic in discussions of responsible conduct is perplexing, considering the extent of attention it commands in the legislative and protest arenas.

Scholarly leadership in the United States has been mobilized to seek experimental alternatives to the uses of animals in toxicity testing. Spearheaded by the Johns Hopkins Center for Alternatives to Animal Testing (CAAT) and the Center for Animals and Public Policy at Tufts University, and with funding contributions from the cosmetic and pharmaceutical industries, a research initiative is in place to identify improved ways of assuring safety and efficacy in products in order to reduce adverse impacts on animals.

Most Americans are unaware of the complexity or dilemmas of animal use. Whether presented by the scientific or animal protection communities in the media or at conferences, information is often delivered in sound bites, oversimplified debates, or pictorial images of extreme cases, rather than in reasoned consideration of the challenging questions. The testing of rabies vaccine for the presence of live virus or for efficacy raises more complex questions than the testing of new shampoos on animals. Young women who use pregnancy test kits generally are not aware that the kit may use monoclonal antibodies produced in mice, a procedure that probably caused the mice pain and suffering—and that an alternative exists.

Legislation and Regulation

In many countries, the regulation and monitoring of animal care and use have begun at academic institutions and subsequently radiated out to include industrial and other research organizations that conduct scientific procedures with animals. Commonly, a national board or panel provides centralized oversight by making institutional visits on a regular basis. Often, a local committee consisting mostly of scientists at the institution focuses on animal ethics or animal care and provides ongoing review of animal use procedures within that institution. Some countries, such as Switzerland, Germany, and the United Kingdom, require a project license for conducting procedures with animals.

United Kingdom

Centralized oversight of animal use is the hallmark of the United Kingdom's legislation and administrative structure. The Home Office provides the governmental oversight and assures accountability of all individual scientists and institutions by requiring licensure (Townsend and Morton, 1995). The Animals (Scientific Procedures) Act of 1986 requires that animal procedures can take place only in designated premises that have been certified to have the necessary facilities to properly care for the animals. A project license is required for the overall program of work, based on a careful weighing of the potential animal suffering against the potential benefits of the research (the cost-benefit analysis); a consideration of available nonanimal methods; a minimum number of animals, especially dogs, cats, and primates; and minimal discomfort or suffering. Severe pain or distress to animals is not permitted. The people who conduct the study are required to obtain a personal license, based on the necessary training, skills, and experience. Violations of personal or project licenses are criminal offenses. The United Kingdom also has an Animal Procedures Committee, a national policy committee that makes recommendations to the secretary of state. It requires that half the members have not been involved in research for the previous six years.

The Netherlands

The Netherlands has been noteworthy for its collaboration between animal rights and laboratory animals communities in developing legislation (Homberger and Thomann, 1996). Requirements in the Netherlands are particularly stringent in comparison with other countries. For example, three weeks of training are required for researchers who will work with animals. The production of monoclonal antibodies within mice has been banned. The 1997 Experiments on Animals Act requires an ethical review of all animal procedures by a committee, the chair and two members of which cannot be employees of the institution. It further prohibits any animal experiments for new or existing cosmetic products and prohibits any median lethal dose (LD_{50}) or median lethal concentration (LC_{50}) tests. The new legislation endorses the concept of the intrinsic value of animals.

European Community

Northern European countries typically evidence great concern with animal welfare, with the result that EC regulations are probably the most stringent in the world (Homberger and Thomann, 1996). Strong leadership in some northern European countries has resulted in the adoption and implementation of strong measures by all participating countries.

Canada

The Canadian Council of Animal Care (CCAC) began as a professional organization serving academic institutions. Still continuing as a professional, voluntary process, the CCAC is well recognized as a progressive and effective system for optimizing animal care.

Australia and New Zealand

Lacking huge cosmetic and pharmaceutical industries, Australia and New Zealand have focused on the uses of animals in research and teaching. Strong and proactive leadership among scientists in Australia and New Zealand has moved laboratory animal care steadily forward. The Australian Code of Practice for the Care and Use of Animals for Scientific Purposes established a process for the ethical review and refinement of animal use protocols. The Australian and New Zealand Council for the Care of Animals in Research and Teaching (ANZCCART) serves as an advisory group, offers regular conferences that involve international participation in discussion and debate concerning ethical and practical aspects of animal welfare, and publishes an excellent newsletter that systematically addresses the care of each laboratory species. Often, initial legislation is passed in the Australian state of New South Wales, and similar legislation develops later within the country. From the beginning, cooperation among the research community, animal welfare groups, and local government has created a forum for people to exchange their views and develop policy. Each ethics committee is required to have at least one public member and one animal welfare member.

The code of New Zealand emphasizes performance standards and covers all vertebrates (Kirton, 1997). Three unaffiliated members required on the institutional animal ethics committees are a lay member, a veterinarian not with the institution, and a representative of an animal welfare organization. No license is required. However, the review process requires researchers to demonstrate the value of the proposed work and the committee to assess the impact on the animal's welfare via a cost-benefit analysis. Researchers must also justify the lack of alternatives.

Japan

Japan has published ethical codes on animal welfare that supercede laws or regulations (Nomura, 1995; Prime Minister's Office, 1980). No inspection system with punitive measures or enforcement is maintained. However, certain laws and standards apply to universities and other research institutions. Most universities, especially medical schools, have an animal experiment committee, which generally reviews protocols.

United States

Legislation for the care and use of laboratory animals in the United States is set forth in the Animal Welfare Act, which includes laws passed by Congress in 1966, 1970,

1976, and 1985 (Hamm, Dell, and Van Sluyters, 1995). The specific regulations that implement the Animal Welfare Act are within the Code of Federal Regulations (USDA 9 CFR, Parts 1–3, 1989). They cover mammals but exclude rats, mice, birds, and farm animals that are not used in research. The U.S. Department of Agriculture (USDA), through inspectors in the Animal and Plant Health Inspection Service (APHIS), Animal Care, administers the Animal Welfare Act. Additional legislation applies to all animal uses conducted or supported by the Public Health Service (PHS). The Health Research Extension Act sets forth the *Public Health Service Policy on Humane Care and Use of Laboratory Animals* (OPRR, 1986), which applies to any live vertebrate animal used in research, training, or testing. The NIH Office for Protection from Research Risks (OPRR) administers the PHS policy, which requires that each institution provide an acceptable assurance of meeting all minimum regulations and conformance with the *Guide for the Care and Use of Laboratory Animals* (National Research Council, 1996) prior to conducting any activity involving animals. The typical method for demonstrating assurance is to seek voluntary accreditation from the Association for Assessment and Accreditation of Laboratory Animal Care (AAALAC) International, which verifies conformance with the *Guide*.

All U.S. universities that are accredited and receive grant support are required to provide equal treatment for all vertebrates. Reviews of specific protocols for fish, reptiles, mice, dogs, or monkeys to be used in research, teaching, or testing are administered through the same process. Anyone using animals in a U.S. university submits a protocol to the institutional animal care and use committee (IACUC). Yet, someone establishing a company to develop new strains of transgenic mice or to produce monoclonal antibodies in mice may not experience this process. For facilities using only rats and mice and not receiving grant support from a federal agency, accreditation by AAALAC International is optional.

The U.S. regulations specify that researchers consider alternatives to "procedures that may cause more than momentary or slight pain or distress (USDA 9 CFR, 1989, p. 36152, paragraph 2.331). Most of the attention on alternatives has focused on uses of animals in testing, stemming from the high public awareness of the Draize test and the LD_{50} test, and is addressed by the efforts of CAAT at Johns Hopkins University. Government involvement was stimulated when the NIH Revitalization Act of 1993 directed the National Institute of Environmental Health Sciences to address the need for validation and regulatory acceptance of toxicity testing methods. Thus in 1994, the Interagency Coordinating Committee on the Validation of Alternative Methods (ICCVAM) was formed.

The biomedical research sphere in the United States is largely driven by creative desires to answer new scientific questions. The vast world of tissue and cell culture, mice and rats, and biotechnology seems remote from the ethologists who study animals in their natural environment and evolution. Yet another area involves creating new products, including pharmaceuticals and vaccines, that must be tested for safety and efficacy. In the United States, these various worlds are linked when they are located on university campuses because all procedures involving animals are reviewed by the same IACUC. This committee is composed of researchers and individuals who are somewhat knowledgeable about science; even the required public member often is a former academic employee who has retired.

In the United States, changes in legislation and regulation regarding animal welfare often are affected by the targeted campaigns of animal rights protests. Similarly, the pharmaceutical and cosmetic industries have responded to the public pressure that has focused on toxicity testing. These industries provide detailed information on animal use to their stockholders. For example, one stockholder in American Home Products proposed at the 1993 annual meeting to take all necessary steps to eliminate all animal testing wherever possible (American Home Products Corporation, 1993). In denying the proposal, the board of directors and management drafted a detailed reply to all stockholders and described spending more than $1 million in the previous two years on developing alternatives to the use of animals in testing. The company had reduced the use of animals in testing by 30 percent over a five-year period. The mouse safety test and the rabbit pyrogen test, two in vivo procedures, had been replaced with in vitro procedures.

Monitoring the Numbers of Animals Used

The concept of alternatives, as described in the three *R*s (Russell and Burch, 1959), has furthered the view that, when possible, reducing the numbers of animals used in scientific procedures is desirable. Many countries maintain and publish detailed censuses of all animals used in various scientific procedures, listed by species and type of procedures. The EC publishes an annual report of animal use and aspires to increase the accuracy of the estimates of animal numbers by its member countries (Straughan, 1994). Over time, it will be possible to closely monitor changes in animal use. This aim for reduction in animal use has been somewhat countered in the United States by the American Medical Association's preference for the term *adjunct* rather than *alternative* (American Medical Association, 1992; Message in a bottle, 1995). Further, the expanding emphasis on uses of transgenic animals to model human diseases and develop new vaccines increases the likelihood of an increase in the numbers of animals used rather than a reduction (Gordon, 1997).

United Kingdom

Each year, a detailed record is available from the United Kingdom that reports the numbers of procedures involving animals, by species and by type of procedure. Animals that are simply euthanized without being subjected to a procedure or are killed for tissue use or samples are not included in the count. All living vertebrates, including certain transgenic animals and genetic mutants, as well as *Octopus vulgaris,* are covered by the legislation. The censuses make it possible to monitor changes in patterns of animal use and to judge the extent of discomfort caused.

The Netherlands

Unlike most other countries, the census of animals in the Netherlands includes animals that were killed without any experimental procedure being performed. Furthermore, the 1996 census revealed that no animals were used to test cosmetics.

United States

The United States monitors six species: dogs, cats, guinea pigs, hamsters, rabbits, and primates (a category that includes numerous species). Except for primates, research, teaching, and testing uses of these species have declined markedly between 1973 and 1994 (Crawford, 1996; Davis, 1996; Lingeman, 1996). In federal research facilities that keep records on rats and mice, use of rats and mice has also declined (Davis, 1996), but figures are not available for the overall use of rats and mice nationally. One credible estimate reported that 11 to 19 million mice and rats were used in 1994 (Rowan and Loew, 1995). Universities generally are accredited by AAALAC, which requires the same oversight for rodents and for the other six species and the maintenance of complete records on all animals. Lacking regulation for rats, mice, and birds, the United States cannot systematically assess its patterns of use for the rodents that now account for about 90 percent of the animals used. This fact places in question the sincerity and integrity of the scientific enterprise in seeking to improve the welfare of laboratory animals.

The number of unregulated and unaccredited animal facilities for rats and mice in the United States is not known. Legislation passed in 1989 by the city of Cambridge, Massachusetts, "Ordinance for the Care and Use of Laboratory Animals in the City of Cambridge," requires that all experiments on all animals be in conformity with the *Guide for the Care and Use of Laboratory Animals* (National Research Council, 1996), the Animal Welfare Act, and the *Public Health Service Policy on Humane Care and Use of Laboratory Animals* (OPRR, 1986), among others (Ordinance no. 1086, Publication no. 2453, June 29, 1989, Cambridge, Mass.). Violation of the ordinance carries a possible fine of $300 per day. In recent years, about five of Cambridge's more than 25 animal facilities became accredited by AAALAC International (J. Medley, City of Cambridge, private communication, 1995). The Cambridge figures illustrate that without the city's legislation, many laboratory rats and mice in the city would have no external oversight. Close monitoring is important because mice are used for some of the most invasive procedures (Hart and Mitchell, 1995). Lacking regulatory control of mice and rats impedes systematic efforts to alleviate their pain and suffering through refinement or other alternatives. The number of animals killed is also a growing concern.

Efforts by animal protection organizations to file a legal suit and gain a judicial hearing on the regulation of rats, mice, and birds have been stymied by the issue of legal standing. Thus, when the Animal Legal Defense Fund and others attempted to sue the USDA for not covering rats, mice, and birds within the regulations, the case was dropped because rats and mice lack legal standing, and the plaintiffs lacked the legal standing to represent rats and mice. As a result, the core issues of the case were not addressed. A similar bottleneck of legal standing has also precluded hearing the arguments concerning the patenting of mice with human genes and basing regulations on performance versus engineering standards.

Funding for Biomedical Research and Animal Welfare

United Kingdom

Although not extensive, some funding from animal protection groups in the United Kingdom has been provided to conduct research on animal welfare. Organizations such as the Royal Society for the Prevention of Cruelty to Animals have, for a number of years, worked closely with the scientific community to further animal welfare and assist in funding graduate students.

The Netherlands

The Netherlands has established the Alternatives to Animal Experiments Platform, jointly sponsored by industries, government ministries, animal protection societies, and private funds and funded at approximately $1.5 million per year (de Greeve and de Leeuw, 1996). The platform has subsidized projects since 1984 and was formally established in 1987. In 1995, 14 projects were funded, including a project for the development of teaching programs within the higher education curriculum.

The European Community

The establishment of the EC directly benefits animals. The member nations have made it a priority to correct the significant inconsistencies among nations in their treatment of animals. They have been sufficiently motivated to develop a substantial budget. Thus, in Europe there are resources available for animal alternatives projects.

United States

Public funding for basic biomedical research in the United States far exceeds that of other countries. For example, in 1997, the U.S. budget for basic biomedical research is almost $13 billion (Cohen, 1996), many times higher than that of other nations: Australia ($62 million), Sweden ($65 million), New Zealand ($13 million), Canada ($182 million), United Kingdom ($438 million), Japan ($1.2 billion), and Germany ($3 billion). The numbers of animals used in the United States probably are also the highest per capita. The mammoth scale of scientific research and development in the United States creates an inertia such that the formation of a national leadership to establish innovations or disseminate model programs regarding animal welfare is more difficult.

The United States has passed laws and regulations that require enforcement without the legislative commitment to sufficiently fund and implement them. Concerns about the lack of funding contribute to the federal agencies' unwillingness to consider the added responsibility that would be involved in regulating rats and mice. Governmental funding agencies support a significant portion of the costs of basic research. The costs of product development, including providing evidence for efficacy and safety as required by federal agencies, is borne by private pharmaceutical, cosmetic, and household product industries, which also fund efforts to develop improvements in toxicity testing.

Setting Goals Concerning Harmonization and Animal Use

Animals are viewed as precious pets in northern Europe, where the animal rights movement has been vocal and, in the case of the United Kingdom, even very violent. Countries in southern Europe more typically take a more utilitarian view of animals. Since the establishment of the EC, the participating countries have made serious efforts to harmonize practices regarding animals and legislation across national boundaries.

European Community

In 1986, the Council of Europe issued the European convention on the protection of all vertebrate animals used in science (ETS 123; Council of Europe, 1986). Almost all European countries have ratified ETS 123 or are members of the European Union (EU). Subsequently, the EU Directive 86/609/EC (European Commission, 1986) was issued to legislate harmonization across national boundaries. The convention and directive require that all experiments that use animals must be evaluated and registered. The directive addresses those procedures involving pain and distress and requires two weeks of training for those who work with vertebrates. In a major move to address toxicity testing, the European Center for the Validation of Alternative Methods (ECVAM) was established and funded with substantial contributions from the member nations. The formation of the EC has led the members of the European Parliament to pass legislation to harmonize treatment of animals across national boundaries in a wide array of situations, including transport and slaughter. Also, the EC has set a target as part of the Fifth Environmental Action Programme; amendment 100 calls for, by the year 2000, a reduction by 50 percent of the 1991 number of vertebrate animals used for scientific purposes in the EU (Matfield, 1997). Certain tests may be phased out. For example, the abnormal toxicity test is being phased out for the safety testing of vaccines, a decision recommended by an ECVAM workshop and then adopted by the Pharmacopoiea Commission (FRAME, 1996). The Federation of European Laboratory Animal Science Associations (FELASA), representing 19 European countries, plays a consulting role, works to harmonize laboratory animal science in Europe, and issues recommendations.

Proactive Responsiveness to the Public

Advances in the care of laboratory animals worldwide have come about from the combined efforts of the animal protection and scientific communities, with varying levels of cooperation between these entities in different countries. Scientists in the United States are relatively inactive in seeking to address public concerns, which has led to recent editorials in *Science* that call for activist scientists (Daie, 1996) and refer to the scientific community as now "desperately seeking friends" (Greenwood, 1996). Although joint cooperative efforts in drafting legislation or conducting research on animal welfare commonly occur in some European countries and Australia, concerns about public disquiet and a need to further involve the public in ethical decisions are expressed in the EC (Clark, 1993; Sandoe, 1993).

In the United Kingdom, FRAME has excelled in developing materials that provide very basic information and curb common confusion concerning animal use. By creating attractive four-page documents on animal experimentation, cosmetics testing alternatives, and alternatives to animal testing, FRAME has made it possible for interested readers to clarify the difference between research and testing and to acquire accurate information. The same cannot be said for an attractive pamphlet prepared by Join Hands (1997), *Alternative Research Methods: Refinement, Reduction, and Replacement of Animals Needed in Scientific Research,* which creates confusion by emphasizing research in its title but then discussing mostly testing.

The IACUCs in the United States are required to include one public member; often this person is an institutional retiree or scientist. Other countries have done much more to seek consensus and understanding with the public while also providing leadership from the scientific community to improve animal care.

Human Factors

Most discussions of animal alternatives have focused on the problems of pain and suffering. However, the requirement to reduce the numbers of animals used implies the possibility that death to animals also may be a harm. This issue deeply disturbs and concerns many veterinary staff whose responsibility it is to euthanize laboratory animals. If death is considered to be a harm, then it would be appropriate to review the rules precluding reuse of laboratory animals and to weigh the cost of pain and suffering for more animals against the deaths of more animals. This issue generally is not being discussed, although it recently was raised as a newsletter target question (SVME Issues Forum, 1997).

An enthusiastic proactive effort by the scientific community to enhance animal care wherever possible would raise the credibility of scientists as caring individuals and temper the effects of some extreme caricatures of scientists. It would also validate the genuine concerns of some laboratory employees, who sometimes feel marginalized when they express sadness for animals. Andrew Rowan often remarks that no one welcomes using animals; Michael Balls describes similar sentiments. This goal of using animals only when doing so is essential will be more believable to the public if a widespread commitment is demonstrated to implementing alternatives to uncomfortable procedures whenever possible.

Special risks, dilemmas, and personal considerations face the people who work with animals. For example, scientists may find that many people have negative opinions about their work with animals, and even their own children may express concern. A scientist also faces the risk that a local columnist will pinpoint the work and portray it on the front page. University students occasionally go so far as to create an effigy or otherwise criticize the work, often with an inaccurate perception of the work.

According to one study, psychologists, while supporting animal studies that involve observation or confinement, disapprove of studies involving pain or death, especially those involving primates or dogs more so than pigeons and rats (Plous, 1996). A majority of respondents in this survey believed that rats and mice were covered under the Animal Welfare Act.

One Japanese tradition honors the contribution of animals that have given their lives on farms, in slaughterhouses, or in laboratories by establishing a monument and holding an annual ceremony. A similar tradition is followed at the Cleveland Medical Clinic and the University at Guelph. This ritual acknowledgment of the animals' contribution provides some comfort to the people who grieve the animals' deaths.

Conclusion

Increasingly, countries are establishing a process for reviewing animal procedures designed to assess and reduce the cost of the procedure to the animal. This responsibility often is assigned to an animal ethics committee and reflects a primary concern with ethical issues. In a number of countries, animal ethics committees achieve a compatible working relationship with strong representation from both the scientific and the animal protection communities. Most important, the EC has been a catalyst and set ambitious goals for the improvement of animal welfare, developed legislative initiatives, and stimulated funding to benefit animal welfare.

References

American Home Products Corporation. Notice of annual meeting and proxy statement. American Home Products Corporation, New York, March 17, 1993.

American Medical Association. *Medical Progress: A Miracle at Risk, Resource Kit.* American Medical Association, Chicago, August 17, 1992.

Anderson, C. Agencies set rules on financial disclosure. *Science* 265: 179–180, 1994.

Bekoff, M. Experimentally induced infanticide: The removal of birds and its ramifications. *Auk* 110: 404–406, 1993.

Clark, J. M. Animals in research—facing up to public disquiet. Fifth FELASA Symposium: Welfare and Science, pp. 328–330, Federation of European Laboratory Animal Science Associations, London: Royal Society of Medicine Press, 1993.

Cohen, J. Research "summit" ponders health funding shortfall. *Science* 274: 491–492, 1996.

Council of Europe. The European convention for the protection of vertebrate animals used for experimental and other scientific purposes. ETS 123. March 10, 1986, Strasbourg, France.

Crawford, R. L. A review of the animal welfare enforcement report data 1973 through 1995. *AWIC Newsletter* 7 (2, Summer): 1–11, 1996.

Daie, J. The activist scientist. *Science* 272: 1081, 1996.

Darwin, C. *The Expression of the Emotions in Man and Animals*. Chicago: University of Chicago Press, [1872] 1965.

Davis, E. *Animal Use Trends in the United States, 1986–1994*. Vienna, Va.: WARDS, Inc., 1996.

Dawkins, M. *Animal Suffering: The Science of Animal Welfare*. London: Chapman and Hall, 1980.

de Greeve, P., and W. de Leeuw. The alternatives to animal experiments platform. *Alternatives to Laboratory Animals* 24 (special issue): 204, 1996.

Donnelley, S. Introduction. The troubled middle *In medias res. Hastings Center Report* (special supplement) May–June: 2–4, 1990.

Emlen, S. T. Ethics and experimentation: Hard choices for the field ornithologist. *Auk* 110: 406–409, 1993.

European Commission. Directive on the approximation of laws, regulations and administrative provisions of the member states regarding the protection of animals used for experimental and other scientific purposes. 86/609/EC. Brussels, November 24, 1986.

Farnsworth, E. J., and J. Rosovsky. The ethics of ecological field experimentation. *Cons. Biol.* 7: 463–472, 1993.

FRAME. Fewer animals needed for vaccine testing. *FRAME News* 39: 5, 1996.

Garner, R. Animal experimentation and pluralist politics. Plenary address, Second World Congress on Alternatives and Animal Use in the Life Sciences. Utrecht, Netherlands, October 20–24, 1997.

Gordon, J. W. Transgenic technology and laboratory animal science. *ILAR Journal* 38(1): 32–41, 1997.

Greenwood, M. R. C. Desperately seeking friends. *Science* 272: 933, 1996.

Guidelines for the treatment of animals in behavioural research and teaching. *Animal Behaviour* 47: 245–250, 1994.

Hamm, T. E. Jr., R. B. Dell, and R. C. Van Sluyters. United States. *ILAR Journal* 37(2): 75–78, 1995.

Hart, L. A. and A. Mitchell. The (almost) all-purpose laboratory mouse. In *Animals in Science Conference: Perspectives on Their Use, Care and Welfare,* N. E. Johnston, ed. Monash, Australia: Monash University, 1995.

Homberger, F. R., and P. E. Thomann. Harmonizations of animal welfare standards in Europe. *Lab Animal* 25(8): 35–38, 1996.

Join Hands. *Alternative Research Methods: Refinement, Reduction, and Replacement of Animals Needed in Scientific Research.* Washington, D.C.: Join Hands, 1997.

Kaiser, J. Commission proposes new definition of misconduct. *Science* 269: 1811, 1995.

Kirton, A. Functioning of a busy New Zealand Animal Ethics Committee. *ANZCCART News* 10(1): 5–6, 1997.

Lingeman, C. H. 1996. Trends in animal use in U.S. biomedical laboratories. *In Vitro Toxicology* 9(1): 19–42, 1996.

Lorenz, K. *Here Am I—Where Are You? The Behavior of the Greylag Goose*. New York: Harcourt Brace Jovanovich, 1991.

Matfield, M. European Parliament attempts to resurrect the 50% target. *EBRA Bulletin* February: 1–2, 1997.

Mayer, F. L., E. A. Whalen, and L. A. Rheins. A regulatory overview of alternatives to animal testing: United States, Europe and Japan. *Journal of Toxicology—Cutaneous and Ocular Toxicology* 13(1): 3–22, 1994.

Message in a bottle. *The Economist* 335 (7911): 83–85, 1995.

National Research Council. *Guide for the Care and Use of Laboratory Animals*. Washington, D.C.: National Academy Press, 1996.

Nomura, T. Japan. *ILAR Journal* 37(2): 60–61, 1995.

Plous, S. Attitudes toward the use of animals in psychological research and education. *American Psychologist* 51: 1167–1180, 1996.

Prime Minister's Office. Standards relating to the care and management of experimental animals, Notification no. 6, Tokyo, March 27, 1980.

Office for Protection from Research Risks. *Public Health Service Policy on Humane Care and Use of Laboratory Animals*. National Institutes of Health, Rockville, Md., 1986.

Rowan, A. N., and F. M. Loew. *The Animal Research Controversy: Protest, Process and Public Policy*. New Grafton, Mass.: Center for Animals and Public Policy, Tufts University School of Veterinary Medicine, 1995.

Russell, W. M. S., and R. L. Burch. *The Principles of Humane Experimental Technique*. London: Methuen, 1959.

Sandoe, P. Involving the public in ethical decisions. Fifth FELASA Symposium: Welfare and Science, pp. 331–334, Federation of European Laboratory Animal Science Associations, London: Royal Society of Medicine Press, 1993.

Schachman, H. K. What is misconduct in science? *Science* 261: 148–149, 183, 1993.

Straughan, D. W. First European Commission report on statistics of animal use. *ATLA* 2: 289–292, 1994.

SVME Issues Forum. *Society for Veterinary Medical Ethics Newsletter* 3(1): 1–9, 1997.

Tinbergen, N. *The Animal in Its World*. Cambridge, Mass.: Harvard University Press, 1972.

Townsend, P., and D. B. Morton. United Kingdom. *ILAR Journal* 37(2): 68–74, 1995.

USDA 9 CFR, Parts 1, 2, and 3. *Federal Register, Rules and Regulations* 54(168), Thursday, August 31, 1989, p. 36152, Institutional Animal Care and Use Committee (IACUC).

2

DONALD A. DEWSBURY

Is the Fox Guarding the Henhouse?

A Historical Study of the Dual Role of the Committee on Animal Research and Ethics

Although it is generally agreed that virtually all professional activities require some kind of regulation, there is disagreement as to how this regulation should be effected. On the one hand, allowing members of a profession to create their own regulatory apparatus ensures that the regulators have the technical expertise to evaluate practices appropriately. Proponents argue that it is in the best interests of all practitioners to weed out those who behave inappropriately and thus protect the reputation of the profession at large. On the other hand, it is argued that regulation should come from outside of the profession so that objective perspectives can prevail because members of the profession will protect each other and effective regulation will be impossible. In the context of animal care issues, this perspective is the spirit behind the policy that institutional animal care and use committees (IACUCs) must have community representatives (Orlans, 1993, chapter 7). Although there are many regulatory agencies outside professional organizations, most such organizations have some kind of internal means of quality control, and such groups must deal with the conflict inherent in the dual role of regulator and practitioner. The medical and legal professions and the U.S. Congress are examples.

What is now called the Committee on Animal Research and Ethics (CARE) of the American Psychological Association has been in existence for 70 years. I will explore some important events in that 70-year history with an emphasis on the duality inherent in the role of regulator and regulated.

The recent activities of CARE often have been intertwined with those of the Psychologists for the Ethical Treatment of Animals (PsyETA), which was founded by Kenneth J. Shapiro and Emmanuel Bernstein in 1981 and incorporated in 1983. According to an information pamphlet produced by that organization:

> Psychologists for the Ethical Treatment of Animals (PsyETA) is an independent association of psychologists dedicated to the promotion of animal welfare within the science and profession of psychology. . . . PsyETA stands for an ethical dimension, balancing the value of experimentation against the suffering of animals. While recognizing the benefits of research, we hold that the rights and interests of the nonhuman animals involved are substantial and must be respected. (PsyETA, 1984a)

This issue of duality came to a head in the recent history of CARE as PsyETA attacked the committee on precisely these grounds. In the words of a PsyETA position paper: "The APA governance should include a committee solely charged with animal welfare. Currently, one committee serves the interests of both the experimenter and his animal subjects. This is a clear case of conflict of interest" (PsyETA, 1984b, p. 1). I shall try to illuminate the history of the committee specifically with regard to this issue.

Two clarifications are in order. First, although the committee has been in continuous existence for 70 years, the name has been changed three times. I will frequently refer to it as "the committee," meaning what is now called CARE. Second, although I refer to them as "regulators," any regulation by the committee has been very soft regulation, as it has had virtually no power to enforce recommendations it might promulgate.

The Founding of the Committee

The founding of the committee is a story of an individual practitioner, in this case a physiologist, who urged the development of standards as a means to protect all practitioners.

The committee was formed at the instigation of the physiologist Walter B. Cannon, who addressed a letter to the American Psychological Association (APA) Council in 1924 (Bentley, 1925; Dewsbury, 1990). Cannon was concerned about an article in the *Journal of Comparative Psychology* by future APA President Calvin P. Stone (1923). The experiment that was reported involved surgery; although it appeared certain that anesthetics were used, Stone failed to mention this fact. In letters to both his friend Robert M. Yerkes (Cannon, 1924a) and journal editor Knight Dunlap, Cannon (1924b) criticized Stone's report, pointing out that the paper left all open to attacks from antivivisectionists. On February 2, 1925, APA President Madison Bentley appointed an ad hoc committee of Yerkes of Yale University as the chair, Paul Thomas Young of the University of Illinois, and Edward C. Tolman of the University of California, Berkeley (Bentley, 1925). This committee remained intact through 1930. Clearly, Cannon's proposal was self-regulation to protect the integrity of those psychologists doing animal research.

The committee recognized that some regulations concerning animal use were needed and, rather than developing its own, simply adopted those of the American Medical Association (AMA). Cards with the AMA Code of Animal Experimentation printed on them were prepared for posting in laboratories of animal psychology. The code covered issues such as the use of vagrant cats and dogs, kind and sanitary care and maintenance, responsible supervision of surgical procedures, appropriate use of anesthesia, and painless euthanasia (Tolman, Young, and Yerkes, 1926).

Five resolutions were adopted by the APA at the instigation of the committee (Young, 1928). The first was the acceptance of the AMA regulations. The second concerned an "open door" policy toward officials of humane societies. The third requested the cooperation of journal editors in enforcing the code for animal experimentation, recommending that editors "decline to publish manuscripts descriptive of experiments which violate the code of animal experimentation adopted by the Association" (p. 489). The fourth resolution concerned classroom use of animals. The final resolution recommended the appointment of a standing committee on precautions in animal experimentation. The tone of the committee activities during these early years was that practitioners were developing standards and policies to protect practitioners.

The Ebb and Flow of Committee Activity

Committee activity ebbed and flowed over the years in association with more global forces both inside and outside psychology. Initially, the committee made a deliberate decision to take a low-key approach: "the Committee should not be aggressive unless an unforeseen crisis should arise" (Tolman, Yerkes, and Young, 1929, p. 131). This approach shifted in response to events in the world at large and to changing environments for animal research. World War II had an enormous impact on the nation and on psychology, as one-quarter of all Americans with graduate degrees in psychology were involved in the military effort. In the context of the war mobilization, animal research seemed irrelevant; committee chair Robert Tryon reflected that "animal experimentation seems now to have an 'other world' character" (Tryon, 1942).

Although the committee did not meet formally during its early years, it later moved to one annual meeting and then shifted to two annual meetings. In 1985, an unprecedented three meetings were held. The increase reflected pressures related to the necessity of dealing with issues raised by members of the animal protection community.

The Committee Name

The committee has existed under four different names. It originated as the Committee on Precautions in Animal Experimentation. There is no evidence that the initial name was the result of any special deliberation; the usage is present in the earliest correspondence about the committee.

In 1960, the committee became the Committee on Precautions and Standards in Animal Experimentation. The reason for the addition of "and Standards" was the decision "that the scope of its area be broadened to include scientific standards in animal research" (Carter, 1960, p. 763). This change was made prior to outside pressures to establish standards.

In 1978, the committee became the Committee on Animal Research and Experimentation (Conger, 1978, p. 565). In recommending the name change in its 1977 report, the committee termed the old name "cumbersome." Adoption of the acronym CARE was clearly intended to carry a message.

The current name, Committee on Animal Research and Ethics, was adopted in 1988. In adopting the new name, the APA Council made it clear that "this action is a change in name only and does not reflect a change in the focus or activities of the Committee" (Fox, 1988, p. 527). However, CARE suggested more than that:

> CARE concluded that this name change is not purely cosmetic and considered the implications of adding "ethics" to its name. CARE concluded that the addition would better reflect the *dual mission of the Committee,* as outlined in its Rule of Council (110-3). Further, the Committee acknowledged that the current name, Committee on Animal Research and Experimentation, is redundant and had been selected to achieve the desired acronym, CARE. (CARE Draft Minutes,[1] February 27–28, 1987, p. 2; emphasis added)

The Charge of the Committee

A controversial issue that is especially revealing with respect to the dual role of the committee was the charge, or mandate, of the committee. The issue has centered on whether the committee functions solely to protect animal researchers from critics and legislation or has the dual function of protecting both researchers and the welfare of their animal subjects.

The Initial Charge

As noted previously, the original committee offered five resolutions at the 1925 APA meeting, including the final one:

> The final resolution recommends that the Association maintain a standing committee on "precautions in animal experimentation" which shall cooperate with other organizations interested in safeguarding animal experimentation, and which shall endeavor with them to disseminate accurate information about animal experimentation and to combat attempts to prevent or restrict it. (Young, 1928, p. 489)

The PsyETA advocates, Field, Shapiro, and Carr (1990) quoted this passage, with a trivial error in quotation, in support of the conclusion that the initial charge was concerned only with protecting research and not with "progressive concern with animal welfare" (p. 2). They wrote: "These charges do not refer to protecting animals or to animal welfare or to the ethics of animal research. Rather, the express charges to the committee were to protect then current practices of animal research and to 'combat' attempts to reform it" (p. 2).

As with many such events, this one can be interpreted in different ways, depending on one's perspective. It is true that the actual charge dealt only with the protection of research. However, it is also true that it was just one of five resolutions adopted. The other four resolutions dealt with issues relevant to "progressive concern with animal welfare": adoption of regulations, the open-door policy, interactions with journal editors to reject papers with questionable methods, and the dissemination of the standards to classrooms and the public. Thus, although it may be technically correct that the initial charge specified only the protection of research, it is deceptive to omit the other four resolutions proffered by the committee and adopted by the APA. Clearly, the committee was concerned with protecting both researchers and animals, albeit the latter was in the service of the former. Individuals with different perspectives will differ as to whether the balance between them was appropriate.

Continuing Perspectives

The pattern of concern for both animal welfare and the protection of research, with the former often driven by the latter, continued. An example can be seen in the 1931 report:

> Finally, may your Committee again request that psychologists report to the Committee all matters which might endanger experimental work with animals, such as impending legislation in one's locality, cruel or inhumane treatment of laboratory animals, criticisms by humane societies, or published material which violates the code of rules adopted by the American Psychological Association. (Heron, Warden, and Tolman, 1931, p. 662)

Clearly, the committee was concerned with both cruelty to animals and impending legislation that might hamper research; both were viewed as threats to research.

Reformulating the Charge

I have been unable to locate a copy of the charge to the committee as it existed in the 1970s in the Rule of Council 110-3.1. The effort to reformulate the charge of the committee began in 1979. According to the committee's annual report:

> CARE was stimulated to consider its charge by a letter from a member of APA. . . . CARE noted that its charge is one primarily of promoting research yet ethical issues (e.g., animal welfare) were becoming a dominant concern. While the two are not seen as in direct conflict, the committee believes that its charge needs careful review. (Overmier, 1979)

Presumably the letter in question was that from Emmanuel Bernstein (1979).

At its first meeting of 1980, the committee proposed a new charge "to include more specifically the function of safeguarding the welfare of animals used in research" (minutes of CARE meeting of February 1980). The new wording appears to continue the then-operative wording in points a and b, to revise point c, and to add a new point, d. The following wording was proposed by CARE and approved by the APA Council of Representatives:

> There shall be a Committee on Animal Research and Experimentation whose responsibility it shall be (a) to establish and maintain cooperative relations with other organizations vitally interested in safeguarding animal experimentation, (b) to disseminate, in cooperation with other organizations, accurate information about animal experimentation, (c) to review the ethics of animal experimentation and to combat attempts to prevent or restrict properly conducted animal research, and (d) to disseminate, in cooperation with other organizations, guidelines protecting the welfare of animals used in research, teaching, and practical applications and to consult on the implementation of these guidelines. (CARE draft minutes of the February 1980 meeting; Abeles, 1984)

At its March 1983 meeting, the committee recommended yet another revision of its charge. The new wording was adopted by the APA Council of Representatives:

> There shall be a Committee on Animal Research and Experimentation whose responsibility it shall be (a) to safeguard responsible animal experimentation and to establish and maintain cooperative relations with other organizations [vitally interested in safeguarding animal experimentation] *sharing common interests,* (b) to disseminate, in cooperation with other organizations, accurate information about animal experimentation, (c) to review the ethics of animal experimentation [to combat attempts to prevent or restrict properly conducted animal research,] and *recommend guidelines for the ethical conduct of research, and appropriate care of animals in research,* and (d) to disseminate, in cooperation with other organizations, guidelines for protecting the welfare of animals used in research, teaching and practical applications, and to consult on the implementation of these guidelines. (Abeles, 1984, p. 633; brackets indicate deletion, underlining indicates addition)

Clearly, these changes represent a softening and balancing of the charge to reflect the committee's dual role. It was noted that "these changes are intended to clarify the *dual role of the committee* in protecting the welfare of animals and promoting behavioral research with animal subjects" (Abeles, 1984, p. 633; emphasis added).

The View of PsyETA

Members of PsyETA have generally been unimpressed with the revisions to the charge of the committee. In the words of Field, Shapiro, and Carr (1990, p. 2):

> The charges of the contemporary CARE Committee had the same substance and almost the same wording until the early 1980's (CARE, 1980). Only at that time and for the first time was the combative and defensive language softened, and language regarding protecting animals through ethics, reviews and dissemination of guidelines added (APA Association Rules, 1989). Historically and currently, the CARE Committee embodies a traditional view of animal researchers that their experimen-

> tal procedures should not be restricted and that humane treatment, while of concern, is only considered within that purview.

Although CARE's charge did not always include the dissemination of guidelines until recently, such dissemination was among the first activities of the committee, and the committee has engaged in such dissemination throughout its history.

As noted in the introduction, a persistent source of disagreement between CARE and PsyETA has concerned the establishment of a separate committee within APA to deal solely with animal welfare.

The issue has been addressed repeatedly by PsyETA and first shows up in the materials at my disposal in the letter of Bernstein (1979) and in the 1980 Annual Report of CARE (P. Adams, 1980), prior to the founding of PsyETA:

> Dr. Emmanuel Bernstein attended the meeting as a guest of the Committee and discussed his efforts to get APA to establish a separate committee primarily concerned with animal welfare. . . . CARE agreed that it felt the majority of Bernstein's concerns were being addressed by CARE and recommended against the establishment of a new committee. (p. 3)

The minutes of various meetings show that PSYeta pressed this issue and CARE defended the status quo at several such meetings. In 1987 CARE was informed that the APA Board of Social and Ethical Responsibility for Psychology (BSERP) "denied Psy-ETA's request for the formation of a new committee to address animal welfare issues. BSERP indicated that CARE adequately was addressing these issues" (CARE draft minutes, September, 1987).

In conclusion, it appears that the initial charge to the committee was written narrowly and did not reflect the actual activity of the committee in working to protect animal welfare. More recently, that charge has been revised, and a balance has been struck between protecting researchers and protecting animals. Members of CARE and PSYeta disagree about the actual functioning of CARE, with the former believing that the committee adequately handles both of its major concerns and the latter believing that animal welfare can be addressed only by an independent committee.

Committee Membership

By my count, there have been 106 appointments to the committee (for a list, see Dewsbury, 1993). With some complications due to multiple terms and inabilities to serve, a total of 100 different individuals have served on the committee. All of these individuals have been animal researchers, which implies that the appointing bodies believe that serving effectively on a committee dealing with issues of animal research requires experience at conducting such research. By contrast, PsyETA members believe that there should be an individual on the committee who is not an animal researcher (Bernstein, 1991). Such individuals serve on institutional committees (IACUCs).

Committee Chairs

In the general pattern of succession during the early years of the committee, with three committee members each serving a three-year staggered term, the senior member of the committee served as chair. However, there were various exceptions. Some additional decision making was required after the committee expanded to six in 1958. The chairperson of the committee has generally been elected by the committee members.

I have located no formal policy to suggest any particular qualifications for a chairperson beyond those of being a committee member. However, there has been some informal feeling that the chair should be able to communicate effectively in representing the committee. In addition, there is some feeling that the chair should be familiar with basic surgical procedures; not all psychologists working with nonhuman animals use surgical procedures. The only statement regarding chairs that I have found is in the 1951 committee report:

> It is also recommended that serious consideration be given to the proposal (by W. D. Neff) that future Committee chairmen (1) should be chosen for a period of several years rather than one, and (2) should be willing to work in close contact with the above-mentioned Society for Medical Research. (*It would help if, in addition, the chairman were himself acquainted at first hand with operative techniques in experimentation.*) (Brown, Riesen, and Keller, 1951, p. 1; emphasis added)

Judson S. Brown had accepted as chair of the committee in 1951 only reluctantly because of his lack of experience with surgery. Again, the issue of competence versus distance is apparent.

Field, Shapiro, and Carr (1990) published an analysis of the research of the chairs of CARE, concluding that their research projects were more invasive than relevant comparison groups. Dewsbury (1993) discusses and analyzes this report in his appendix 3, where he concludes that the report has several serious technical flaws: problems of reliability, lack of demonstrated validity, treatment of ordinal data as if they were interval data, use of multiple *t*-tests without adjusting significance levels, unwarranted assumptions of independence in the data set, failure to use blind scoring procedures, and a questionable selection of comparison groups.

Although the analysis of Field and colleagues is flawed, the conclusion appears to be generally valid. Many of the CARE chairs have been physiological psychologists or students of the processes involved in animal learning, especially during the 1980s. There are several possible reasons. First, these areas have become numerically dominant in animal psychology, and thus it would appear likely that more individuals would be chosen from this group. Second, individuals experienced with using invasive procedures may have expertise useful in evaluating research involving surgical and other procedures. Third, individuals using these procedures may be stimulated to participate in CARE activities to a greater extent than others, either because of genuine concern for their animal subjects or because of outside pressures stemming from criticism of their research. An alternative interpretation is that there has been a deliberate effort to select chairs doing invasive research. I find no evidence to support the latter conclusion.

Guidelines for the Use of Animals

Although they did not appear in the formal charge until relatively recently, developing guidelines for the use of animals has been a major activity of the committee since its inception. The initial set of rules was adopted in 1925. Revisions were adopted in 1949, 1962, 1968, 1972, 1979, 1985, and 1992.

In general, these guidelines have become more elaborate and detailed with successive revisions. The initial AMA Rules Regarding Animals, adopted in 1925, contained just five principles. A new set of rules, apparently the first generated by APA itself, was approved on September 8, 1949, by the Council of Representatives upon recommendation from the committee. There were six articles (see Neff, Keller, and Bruce, 1949). Successive revisions altered wording, changed the identity of the individual to whom violations should be reported, and expanded the range of matters covered.

As pressures from animal activists mounted, the 1979 guidelines were the first to contain a preamble, which justified animal research and emphasized the importance of humane care:

> An investigator of animal behavior strives to advance our understanding of basic behavioral principles and to contribute to the improvement of human health and welfare. In seeking these ends, the investigator should ensure the welfare of the animals and should treat them humanely. Laws and regulations notwithstanding, the animal's immediate protection depends upon the scientist's own conscience. For this reason, the American Psychological Association has adopted the following Principles to guide individuals in their use of animals in research, teaching, and practical applications. All research conducted by members of the American Psychological Association or published in its journals must conform to these Principles. (Principles, 1979)

An event that changed the committee's guidelines occurred in 1981, when the APA first incorporated a principle concerning animal use, Principle 10, into its general ethical code that applies to all psychologists. The primary responsibility for ethical questions in animal research was thereby shifted from the committee to the APA Committee on Scientific and Professional Ethics and Conduct (CSPEC), now the Ethics Committee. The ethical code was revised and expanded several times, most notably in 1992 (Ethical principles of psychologists and code of conduct, 1992). The 1992 principle dealing with animal research has nine sections. In brief, they concern (1) humane treatment; (2) lawful acquisition, maintenance, and disposal of animals; (3) proper supervision of procedures; (4) proper instruction of assistants; (5) matching of responsibilities to competencies; (6) minimization of pain and discomfort; (7) consideration of alternatives prior to painful procedures; (8) use of anesthetics during surgery and proper postoperative care; and (9) humane euthanasia. Although the committee never had any real power of enforcement, there was at least some possibility of committee action. Alleged violations that could not be resolved at the local level were to be referred to the committee. The development of principle 10, however, clearly placed responsibility for enforcement elsewhere. Any power would be further minimized as governmental regulations were enlarged and developed and thus came to overshadow those of professional and scientific organizations.

Returning to CARE, the 1985 *Guidelines for Ethical Conduct in the Care and Use of Animals* (Guidelines, 1985) represent a complete restructuring of the earlier principles. A study was made of all the guidelines of various other societies and sources, and an effort was made to address in some way each of the significant issues raised in these documents. A much more elaborate set of principles, with nine major sections, was issued. These guidelines were mentioned favorably in a report on alternatives to animals in research issued by the U.S. Office of Technology Assessment (Fisher, 1986), who called it "the most comprehensive document of this type" (p. 16). However, PsyETA leader Emmanuel Bernstein was quoted as saying that the new guidelines are "merely an elaboration on principles which have existed for more than 30 years with virtually no impact whatsoever" (McCormick, 1985).

The 1992 revision simplified some of these principles and altered them to comply with federal legislation that had been passed in the interim.

Other Activities

The committee has engaged in many other activities in its dual capacity of protecting both animals and animal researchers. There have been extensive interactions with other organizations, including those generally oriented toward protecting animal research and those focused on animal welfare. Extensive campaigns have been launched to publicize the value of animal research in psychology, to provide examples of significant work, and generally to mobilize support within and outside psychology. In recent years, the committee and the APA in general have been increasingly active in lobbying related to congressional activity concerning animal research. It generally has lobbied for weaker government regulation. The committee has frequently sponsored presentations at the annual APA conventions. Early in 1991, the APA Science Directorate established an Animal Research Information Board, an electronic bulletin board maintained by the APA research ethics officer to keep members informed of such issues as activities in Congress, federal regulations, groups for and against animal research, and campus activities. In establishing guidelines for the use of animals in educational settings, CARE has led. The committee has conducted a number of surveys of animal use in psychology. Again, we see both sides of the committee's mandate, as these activities include both those involving standards and regulation and those promulgating animal research, perhaps with a stronger emphasis on the latter.

Conclusion

The APA Committee on Animal Research and Ethics is one of the longest standing APA committees and one of the oldest animal research committees of any organization. It was originally conceived as functioning with the dual responsibilities of establishing humane standards for animal use and promoting animal research in psychology. Throughout its history, it has been a committee of animal researchers working to ensure responsible humane standards, but doing so in the interest of ensuring that animal research would continue. Critics contend that the vested interests of committee members prevent them from serving in an unbiased manner and that a separate committee on ethics is needed. Defenders of the committee argue that only those knowledgeable in the field can fully appreciate and understand the subtleties of animal research. Like lawyers, physicians, and others, interesting conflicts arise when practitioners function as regulators. Whether accurate or not, the committee is vulnerable to the suggestion that the fox is guarding the henhouse.

ACKNOWLEDGMENTS This article is based on material that will be available in Dewsbury, D. A. *A Documentary History of the Committee on Animal Research and Ethics (CARE) of the American Psychological Association (APA)*. Available by request from the Science Directorate, APA, 750 First Street NE, Washington, D.C. 20002–4242.

Note

1. Copies of the draft minutes of the Committee on Animal Research and Ethics are available from the Science Directorate, APA, 750 First Street, NE, Washington, D.C. 20002-4242.

References

Abeles, N. Proceedings of the American Psychological Association, Incorporated for the year 1983. *Am. Psychol.* 39: 604–638, 1984.

Adams, P. Committee on Animal Research and Ethics Annual Report. Washington, D.C.: American Psychological Association, 1980 (unpublished).

Bentley, M. Memorandum upon a committee to consider animal experimentation. (Available in the Yale University Library, New Haven, Conn.) Feb. 2, 1925.

Bernstein, E. Letter to the Committee on Animal Research and Experimentation, June 4, 1979.

Bernstein, E. Letter to Donald A. Dewsbury, June 29, 1991.

Brown, J. S., A. H. Riesen, and F. S. Keller. Report of the Committee on Precautions in Animal Experimentation. Archives of the American Psychological Association, Library of Congress, Washington, D.C., 1951.

Cannon, W. B. Letter to Robert M. Yerkes. (Available in the Yale University Library, New Haven, Conn.) Jan. 22, 1924a.

Cannon, W. B. Letter to Knight Dunlap. (Available in the library of Harvard University, Cambridge, Mass.) Feb. 16, 1924b.

Carter, L. F. Proceedings of the sixty-eighth annual business meeting of the American Psychological Association, Inc. *Am. Psychol.* 15: 750–766, 1960.

Conger, J. J. Proceedings of the American Psychological Association, Inc., for the year 1977. *Am. Psychol.* 33: 544–572, 1978.

Dewsbury, D. A. Early interactions between animal psychologists and animal activists and the founding of the APA Committee on Precautions in Animal Experimentation. *Am. Psychol.* 45: 315–327, 1990.

Dewsbury, D. A. *A Documentary History of the Committee on Animal Research and Ethics (CARE) of the American Psychological Association (APA)*. Available from the Science Directorate, American Psychological Association, Washington, D.C., 1993.

Ethical principles of psychologists and code of conduct. *Am. Psychol.* 47: 1597–1611, 1992.

Field, P., K. J. Shapiro, and J. Carr. Invasiveness of experiments conducted by leaders of psychology's animal research committee (CARE). *PsyETA Bull.* 10(1): 1–10, 1990.

Fisher, K. Animal research: Few alternatives seen for behavioral studies. *APA Monitor* 17(3): 16–17, 1986.

Fox, R. E. Proceedings of the American Psychological Association, Inc., for the year 1987. *Am. Psychol.* 43: 508–531, 1988.

Guidelines for Ethical Conduct in the Care and Use of Animals. Washington, D.C.: American Psychological Association, 1985.

Heron, W. T., C. J. Warden, and E. C. Tolman. Report to the American Psychological Association of the Committee on Precautions in Animal Experimentation. *Psychol. Bull.* 28: 660–662, 1931.

McCormick, C. L. The controversy over brutality to animals. *Freedom* November: 4–6, 1985.

Neff, W. D., F. S. Keller, and R. H. Bruce. Report to the American Psychological Association of the Committee on Precautions in Animal Experimentation. *Am. Psychol.* 4: 476–477, 1949.

Orlans, F. B. *In the Name of Science: Issues in Responsible Animal Experimentation.* New York: Oxford University Press, 1993.

Overmier, J. B. Committee on Animal Research and Ethics Annual Report. Washington, D.C.: American Psychological Association, 1979 (unpublished).

Principles for the Care and Use of Animals. Washington, D.C.: American Psychological Association, 1979.

PsyETA (Psychologists for the Ethical Treatment of Animals). *People for the Ethical Treatment of Animals.* Lewiston, Maine: PsyETA, 1984a.

PsyETA. *Position Paper: Governance and Guidelines.* Lewiston, Maine: PsyETA, 1984b.

Stone, C. P. Further study of sensory functions in the activation of sexual behavior in the young male albino rat. *J. Comp. Psychol.* 3: 469–473, 1923.

Tolman, E C., R. M. Yerkes, and P. T. Young. Report of the Committee on Precautions in Animal Experimentation. *Psychol. Bull.* 26: 130–131, 1929.

Tolman, E. C., P. T. Young, and R. M. Yerkes. Report of the Committee on Precautions in Animal Experimentation. *Psychol. Bull.* 23: 124–126, 1926.

Tryon, R. C. Letter to Willard C. Olson. American Psychological Association Archives, Library of Congress, Sept. 22, 1942.

Young, P. T. Precautions in animal experimentation. *Psychol. Bull.* 25: 487–489, 1928.

3

JOHN P. GLUCK

Change During a Life in Animal Research

The Loss and Regaining of Ambivalence

No matter thinks; every soul of beast is matter; therefore no beast thinks.

—Rene Descartes

A man's profession only enters into the drama of his life when it comes in conflict with his nature.

—George Bernard Shaw

There are reasons to believe that the quality of the philosophical debate on the use of animals in biomedical research, testing, and education continues to improve. The intense modern analysis set off by scholars like Peter Singer, Tom Regan, and Bernard Rollin has been taken up and expanded by a host of philosophers and social scientists who bring added expertise and experience with concepts and methods crucial to the dialogue, such as the issue of personhood (Dennett, 1978), the nature of moral standing (e.g., Beauchamp, 1992), and cross-cultural perspectives (Gluck, Eldridge, and McIver, 1993). Others such as Plous (1991) have worked to dispel simplistic and distorted descriptions of those individuals actively involved in the debate. In addition, the work of cognitive psychologists and ethologists continues to provide valuable and often surprising research data on the mental life of an expanding number of different animal types (see Bekoff, 1994; Cheney and Seyfarth, 1990; Griffin, 1992).

Yet, in the face of these advances, major impediments to rational discourse remain. For example, many positional advocates continue to take basically warlike postures and prefer to fight for and manipulate public opinion by whatever rhetorical device they can muster (Gluck and Kubacki, 1991). Other influential researchers have given blanket assurances either that no ethical issues are, in fact, relevant to the animal debate or that all is certainly well behind the laboratory doors with respect to both research practices and compliance with federal and state regulation. On another side, one could get the impression from some writers that primitive and crude eighteenth-century vivisection is still the best representative model of current animal research activities. Thus, in the face of broadening and increasingly sophisticated empirical and philosophical analysis, the polarities still tend to hold center stage (see Orlans, 1993).

The Troubled Middle

The popularity of extreme polar positions is informative for other reasons. It would be wrong to assume that they are always just artifacts of personally motivated intransigence. Rather, their existence suggests and is testimony to the fact that it may be more comfortable there. After all, those are the places where certainty abounds. Nothing seems more certain than Carl Cohen's (1986) emphatic and unhesitating declaration that animals are in no way part of the moral community and that's that. Even though the justification on closer inspection may appear shaky, the clarity of the assertion helps some of us relax. Similarly, when Tom Regan (1983) asserts that absolute vulnerability requires absolute protection, some of us no longer need to question where we stand and what needs to be done. This comfort in the extremes makes the search for what has been called "the troubled middle" so elusive (Donnelley and Nolan, 1990). We have learned a great deal from proponents of extreme positions, but the returns are clearly diminishing at this time. Continued focus on these positions may, in fact, impede change because they offer little or no quarter for movement. In addition, the middle is where, in reality, all the tension is, and that it is why it is troubled.

The middle position articulates the hardest questions. It is where rights and obligations coexist with issues of utilization. It acknowledges that animals require ethical consideration but is unsure about the limits and boundaries of that consideration. The middle reflects the tension that is seen everywhere and every day where animals are concerned. The middle is where awe, love, and respect for animals stand side by side with gargantuan neglect, brutality, and indifference. It is the domain that must be worked through, the meshing of conflicting opposites. As long as we occupy the edges of the argument, we are not faced with this reality (Gluck and Kubacki, 1991).

But certainly, some individual participants in this debate try to live in this middle ground, and surely extreme positional alliance is not altogether, like species membership, immutable and fixed. We must seek out these individuals. It is time to turn our attention in earnest toward the middle. From the philosophical perspective, the elaborators of the center must be given the lion's share of the attention, as Tom Regan, Peter Singer, and Carl Cohen give space. For example, the nonspeciesist utilitarianism of Raymond Frey (1987) and the analysis of the relationship between obligations and rights and the graded concept of moral standing by Tom Beauchamp (1992) require our attention and analysis now more than ever.

Histories of Change

Given the nature of this debate, there is something important to be gained from looking at decisional histories of individual scientists and researchers who have moved away from polar points of view. Their stories, if told honestly and not just to create an appealing story line, might reveal empirical, philosophical, or cognitive windows of permeability that are crucial for the facilitation of perspective shift and alteration, so desperately needed in this acrimonious debate. A phenomenological analysis might also reveal misunderstandings and confusions that are in need of clarification. Perhaps most important, they would contribute journals of ethical struggle and analysis of individuals open to change or at least not actively opposed to it.

Unfortunately, very few in-depth descriptions of positional modification in the literature are specifically relevant to the animal debate. Modern exceptions include the writings of Donald Barnes, Richard Ryder, Charles McArdle, and Harold Herzog. Except perhaps for Herzog, the others represent shifts from one pole to another. My position is that I continue to believe that there are reasonable and justifiable uses of animals in research. Yet, those justifiable uses are far fewer than I would have originally believed. Further, I believe that too few researchers take seriously the ethical questions posed by an analysis of the animal research enterprise. Rather, ethical issues are often conflated with federal regulatory requirements and purely pragmatic considerations relevant to public relations and image. It is to this body of literature that I devote this chapter.

A Personal History of Change

Childhood and Animals

Although I grew up in the urban environment of New York City, dominated as it is by humans and machines, my interactions with animals, whether pleasant or negative, have always felt important. Many of my primary recollections of childhood involve the troubled and confused feelings inherent in most human-animal relationships. As a boy I remember vividly watching the junk man driving slowly through the back alley by my house in a rickety wooden wagon pulled by a tired-looking, old, dark brown horse. My grandmother encouraged me to show simple respect and acts of kindness to this horse. She seemed to see the ragman and horse as sharing a hard and weary marginal life, not unlike her own, deserving of respite and relief. It was fun to see the horse, something different and out of the ordinary. Yet watching the shiver of his flanks in response to the biting flies, the white sweat around the leather harness, and the protruding ribs left me uneasy because I was certain that he was aware of his hard circumstances. As my grandmother chatted with the man and offered refreshment, I fed the horse vegetables and sugar cubes. These acts came easily to me, even though I was intimidated by this nonmotorized entity. The kindnesses seemed natural and right. I do not remember exactly when the horse stopped coming down the alley, but I remember missing him and looking for him for many months, with carrots and sugar at the ready.

I also recall that my grandmother's dog, a white spitz named Buddy, and I got into a hassle over the ownership and control of a gardening tool when I was quite young. The dog bit my right wrist and held me in his grip until my frightened and indignant cries brought rescue. If my recollection is correct, after a proper investigation of the facts, I was punished and the dog was given something special to eat. It would appear that Buddy had a level of standing in my household and possessed a point of view that required consideration. This experience apparently did not poison my relationships with pets, as my mother tells me that I was inconsolably sad when my pet cocker spaniel died some years later. In general, the illness of death of a pet in my home was not a casual event. The tears and upset were obvious in everyone and expressed without embarrassment, and the small cash reserves were expended willingly and without question. In these days, both physicians and veterinarians made house calls, so medical treatment of the patient frequently took place in the home. I can still see our veterinarian's 1949 gray Ford parked in back of the house, a sign of both hope and despair.

Then there were also the highly charged love-hate relationships with the local "wild" animals, the squirrels. Although there was not much room to plant gardens in my neighborhood, my family cultivated a small scrap of soil at the rear of the house. In this small parcel, roses and many other brightly colored flowers surrounded a tiny spot of carefully tended grass. A problem arose each spring and summer as my mother and grandmother fought with the local gray squirrels over control of that speck of dirt. The squirrels preferred to dig holes in it to retrieve or bury things, while my family preferred undisturbed flower beds. There seemed to be no middle ground. To hear the angry talk was to anticipate the killing and torture of these trouble-making intruders. It would not have surprised me to see their skins hanging on the clothesline, much the way I have seen cattle ranchers around southwest Texas hang the carcasses of coyotes on boundary fences, as some form of lesson to others who might "trespass." In reality, nothing like that took place. It was purely a war of words. When winter and the harsh weather arrived, the talk changed dramatically to expressions of concern for these and other animals. The effect of the snow and ice storms on their well-being was the topic of conversation. Now, carefully provided winter rations were substituted for insults and threats of murder.

As I recall these memories, it is obvious that I grew up believing animal life has considerable value. Their needs were considered relevant, though certainly not definitive. The message seemed to be that we did not know quite what to do with animals. It was not quite clear where they fit. I initially carried these values of respect and uncertainty with me into my scientific life with animals. These values had the effect of creating a distinct sense of ambivalence that I felt when I first was encouraged to utilize animal lives in research—ambivalent because, though I felt that the science was interesting and important, I could not conclude that the nature of the animal involvement was beyond consideration. I believed that because some of the experimental procedures caused pain, they were therefore not neutral acts. They required justification, and the justification was not a black-and-white affair. It is my belief that this sense of ambivalence is crucial to our operation as ethically attuned and responsible researchers. Both the conduct of research and noninterference in the lives of animals arise from positive value positions. Both positions are defensible and carry considerable ethical weight, and they must be resolved with reference to specific purposes and real contexts. Yet, much transpires in the process of research and science training that facilitates the elimination of that ambivalence—that just does away with the conflict. At that point, the heavy responsibility inherent in research is betrayed.

Into the Valley

My first active incursion into the life of a living animal in the name of science took place in the context of a field biology course and illustrates my ambivalence. In that course, we were learning to set live traps for small wild rodents that would then be turned into museum collections for the purpose of determining species membership. When I checked my traps one morning, I was surprised to find that a small tan mouse had been captured in one of them. The mouse sat huddled near the back of the wire trap surrounded by the cache of uneaten seeds that had lured him. I began to wonder how this animal was to be converted from its present state into a museum specimen. I knew, of course. A sense of interest and uneasiness began to rise simultaneously. Soon the lab instructor described the manner in which this transformation was to be achieved. First, the animal had to be killed. He described the method. He said it would be simple. All we had to do was to pick the animal up by its tail and place it in the palm of the hand, with our fingers securely wrapped around its torso and its head protruding above this grasp. We were then instructed to compress the animal's chest to prevent its breathing. The rationale for this procedure had to do with simplicity, inexpensiveness, and efficient specimen preservation. I was very reluctant, but in the atmosphere of educational sanction I proceeded.

The feel of the small animal was distinctive. I cannot forget it even now, more than 30 years later. Its fur was very smooth, like a broad, thin flake of soap. As I compressed the thorax, the mouse squirmed at first. So I squeezed harder. I could feel the pounding of its heart against the inside surface of my fingers. It was like a small, powerful, pea-sized piston. I was surprised and bothered by its persistence. I had not anticipated how hard it was to kill a healthy animal only a tiny fraction of my size.

Later, a bit shaken, I attempted to speak to the lab instructor about the justification of both the euthanasia practice and the need for all of us in the class to create our own mammal collections. The reality of the mouse and its response to the procedure led to questions about the project in general. During the discussion, he assured me, in an irritated and dismissive tone, that the number of animals collected in this manner and for this purpose in no way threatened the viability of the populations from which the animals were sampled. Hearing that helped. Yet the ambivalence I felt focused my attention beyond these generalized considerations to the specific events themselves. The particular mouse and the way I interacted with it in real time also mattered. The mouse was more than just a numerically insignificant member of an abstract population. The ambivalence formed a second level of guidance. There were the research purpose in general and the details of the interaction with the animals necessary to achieve the purpose. Both required review, and each influenced the justification of the other.

But as the months and years progressed, I found that this instructive ambivalence became quieter. Eventually, this feeling of ambivalence became so quiet that it no longer was available to motivate pause and reflection about the particular experiments I contemplated and carried out. Only the scientific purpose and context of the experiment were salient. In a way, the animals disappeared along with ambivalence.

The Destruction of Ambivalence

Our relationships with animals contain a fundamental ambivalence. I say *fundamental* purposely. I use the word to underline the idea that the push and pull, the awe and fear, and the respect and the contempt of the other are psychologically congenital where animals are concerned. It is therefore my belief that in the case of scientific training and socialization we have tended to evolve toward the development of processes that crush or hide this ambivalence. Unworked ambivalence is difficult to deal with. It leads to feelings of confusion and helplessness and feeds an internal sense of chaos and distress. How we learn to deal with these feelings has enormous implications for who we conduct ourselves as scientists. The shortcut way to deal with these unsettling experiences is to crush, hide, and deny—that is, to create a create a world of apparent or, in the case of the scientific view, potentially manageable and certain dimensions. Therefore, we might expect that the attempts to neutralize this ambivalence would be particularly strong in biomedical science.

Including my undergraduate and graduate education years, I have spent nearly 30 years of my life associated with animal research. That life has involved participation in sometimes quite invasive research that no doubt produced significant suffering in some experimental animals. Without for the moment considering the scientific justification of the research, the question is still, Where was the struggle? What happened to those confused and uncertain feelings that marked my understanding about the place of animals in the world? Why did the hesitancy disappear so quickly?

I have worked in small, minimally funded laboratories and large interdisciplinary research centers with sizable federally funded budgets. As a graduate student, I worked at the Harry F. Harlow Primate Laboratory, held by some members of the animal protection movement to be the standard of research cruelty. Yet, I believe the majority of my ambivalence was already gone by the time I went to graduate school. I have worked with and observed animal researchers from many different disciplines, such as psychology, physiology, field biology, and neurobiology. My laboratory experience is broad and extensive.

As I examine these experiences, I find that certain contextual features were common to all those circumstances that served to minimize the felt sense of ambivalence. Taken together, these features solidified my belief that my research work and that of my colleagues were critically important and should proceed without question. In the following sections, I review several prominent contributions to the diminishing of ambivalence that I can identify in my own life in animal research. Specifically, I address general characteristics of the laboratory environment such as the actual physical attributes, the sense of urgency and personal entitlement produced by the research enterprise, emotional reactions to the use of animals as a hindrance to research, and the organizing idea of research as a type of war. I try to show, with respect to my own experience, that these factors operated to limit the essential ambivalence necessary to keep alive the assessment of the importance of a particular research project that involves incursions into the life of a nonhuman animal. I then trace the return of that ambivalence and the factors associated with the reemergence.

The Laboratory Environment

Like many students of comparative and physiological psychology and related disciplines, my formal training in animal research began early in my undergraduate education in small, ill-equipped rooms in the basements of multiple-use university buildings. In those days, the expanding interest in animal research exceeded the availability of appropriately designed and designated space. Laboratories were frequently established in opportunistically acquired space.

The animal rooms that were close by often housed a variety of species, including monkeys, domestic cats, pigeons, and the typical laboratory rats and mice. All the animals were kept in the same rooms and were not separated by species. Space limitations neutralized any concern for cross-contamination. The cages of the large animals were constructed of galvanized expanded metal that were notorious for sharp edges that cut animals and students alike. As the cages aged and the galvanized surfaces wore away, they became covered with red-brown rust stains, which gave them a permanently dirty and dingy look, even when they were clean. This appearance helped deplete the motivation to keep them as clean as they should have been. The air-handling systems in the rooms were underpowered for the space and the types of occupants, so the rooms were damp and very smelly. In fact, the process of becoming adapted to the smells of the rooms was part of the ritual of commitment. The physical environment was not what it should have been. It should have been cleaner, better designed, and safer for humans as well as for animals. But these less than optimal conditions were rarely discussed and set certain attitudes in motion. The deficiencies became part of the drama and the atmosphere of discovery. I learned I had to look past the dismal physical elements of the laboratory and to center on the ideas. The physical attributes lost their evaluative relevance, not worthy of much concern or attention. The animals were also part of those physical elements.

The care of the menagerie of animals was left to student volunteers and part-time employed graduate students, none of whom had any formal training in animal care. I landed one of the paid positions as an undergraduate and felt honored by the employment. I found that animal care instructions were few and limited to disinfectant mixing instructions, the location of the various animals' feeds, and the feeding time schedule. Equipment was sparse but always included one pair of stained, brown, thick leather gloves and a catching net to be used in case of monkey escapes. I recall no regular or even occasional access by a staff member or a consulting veterinarian. Only the names and phone numbers of the various professors and graduate students were available in the case of animal emergency. Animal diseases were treated in a standard and ubiquitous way. Rodents that appeared ill had tetracycline added to their water, which produced an attractive pink-red color. Monkeys, cats, and dogs appearing lethargic got shots of penicillin. If the animals failed to recover, their loss to the experiment was lamented but not any suffering they might have endured. The lack of veterinary support was a function of both the austere financial situation of some of the laboratories and the confidence of investigators in their knowledge of animal disease and treatment. I was to see this latter attitude expressed over and over again, even in the most financially fit of research institutions. It appeared that they just did not need to know very much about the animals with whom they worked. What mattered was whether they cooperated with experimental tasks. The attitude seemed to express a simultaneous feeling of independence and broad competence.

When a veterinarian is not available or present in a laboratory, more is lost than simply specific medical expertise. The clinical relationship with the individual animal as the focus is also absent. The presence of a veterinarian illustrates that interactions with animals require specialized education and training. Watching as a veterinarian uses the subtlest of cues to assess discomfort and disease serves as an essential reminder that the animal is, in fact, capable of feeling pain, something that is obviously very easy to ignore (Rollin, 1989).

The Sense of Importance and Entitlement

Since the research space was located in a multiuse classroom building, there was a great deal of ambient noise during an average school day. Therefore, laboratory personnel tended to start work late in the afternoon, when student and faculty traffic was low, and finish early in the morning. As any individual who has chosen or been required to work the night shift knows, there can be a special consciousness that goes with working when most others are asleep. Late-night work in the lab added to the sense of importance. It was like being in a very special club. The discussions that took place between students and faculty during those hours were deep and long and took place in the all-night restaurants. The almost clandestine atmosphere added to the sense of task importance. The discussions were about competing theories and how they might be pitted against one another experimentally.

The experiments conducted in those rooms that I can remember varied in ambition and ranged from such topics as maze learning, rodent reproductive behavior, and predator-prey interactions of falcons to technically demanding neurobehavioral experiments such as determining the effect of suction-produced lesions of the medial and lateral hypothalamus on feeding and drinking activities. In general, these experiments were conducted in the context of great curiosity and anticipation and were relatively unsupervised. Research techniques, including surgical techniques, were at times taught directly by faculty and advanced graduate students and often by one partially trained student to another. The idea seemed to be to do nothing to interfere with the process of research creativity but rather to extravagantly promote it. There was great encouragement—even an explicit demand or obligation—to "just do something," "get started," "manipulate something and observe."

This message seemed to hold even if our ideas were inexperienced and naive.

After all, scientists were commonly extolling the virtues of serendipity in the process of scientific discovery. This encouragement in the face of inexperience communicated to me a great sense of entitlement. Identification with the process of scientific research sanctioned me to do things that others not so identified were prohibited from doing. The right to conduct surgeries, administer drugs, alter life histories, test capacities, and even have keys to locked parts of the buildings seemed to come with this identification. In a sense, it all came too easily. But at the time, it led to a sense of confidence in the importance of our experimental ideas. These messages were subtle in the sense that no one declared the existence of these privileges outright, yet they were pervasive. The research hypotheses, the experiments, and the real or potential findings were crucially important, vastly more important than the animals. The animals need not be considered.

Emotion as a Hindrance to Research

Emotional reactions to the fate of animals in research did come up from time to time. Discussion might be occasioned by the experience of a surgical complication, disease, peculiar drug reaction, or perhaps a new student unprepared for the laboratory environment. However they arose, it was absolutely clear that they were not welcome discussions. Crude and apparently uncaring investigators were also occasionally identified and were the subject of ridicule. However, there was a tendency for these discussions not to proceed to the point of questioning the justification of animal costs in research. For example, the focus of the uncaring investigator's ridicule was the poor quality of the resulting science and not the unnecessary cruelty. In contrast, feelings were likely to be cast as a problem in their own right and not an indicator for continued reflection.

I recall a conversation I had about the use of dogs in various types of experimental protocols. In that conversation, my colleague acknowledged that his personal affection for and sensitivity to dogs absolutely prevented him from ever using them in his own research. At the same time, he expressed a feeling of relief in the knowledge that there were other researchers who were not so sensitive and were able to carry on and use dogs in research. In other words, he placed no normative significance whatsoever on his deeply felt sensitivity and discomfort. I did not get the idea that his emotional self-refusal was based upon a considered rejection of philosophical intuitionism. Rather, he saw his personal reactions as an aberration and unfortunate barrier to his scientific undertakings.

I heard a similar position expressed by a prominent experimental psychologist and radical behaviorist, the late B. F. Skinner, in another personal exchange. I had written to Skinner in the fall of 1974 to inform him that a former student of his and friend of mine had died tragically. In his response to my letter, he warmly recounted his memories of the man and some of the scientific achievements that the two of them had accomplished in their work together. Among the achievements he mentioned was the development of an apparatus called the "shock scrambler." They designed this apparatus to reliably deliver specified electric shocks to the feet of rats in learning experiments in a way that prevented them from adopting behavioral strategies that would permit them to escape or minimize the level of the intended shocks. Skinner went on to reveal that because neither of them liked shocking rats, they unfortunately missed making many very important discoveries." Again, sensitivity to animal pain is openly lamented as an impediment to research.

In a recent paper, Herzog (1993) describes a number of graduate student experiences and quandaries that have maintained a certain ethical potency for him throughout his professional life. These quandaries seem to have mitigated the development of polarized positions. For example, he describes that early on in his career an ecological chemist assigned him the task of converting a series of animals into molecular solutions. This task required putting apparently unanesthetized animals into near boiling water. He recounts in his paper that he had little difficulty boiling earthworms and crickets but began to feel viscerally based objections when he proceeded with the scorpions, lizards, snakes, and mice. He tells us explicitly that he could not bring himself to boil live mice. Of particular relevance to our present discussion was his belief that, once his reluctance to boil the mice was discovered by his superiors, they were likely to react by ending his graduate career! In addition, he portrays the personality changes he observed in another graduate student who was doing neurobiological studies with domestic cats that required that, following a series of behavioral experiments, they be killed and their brains fixed and sectioned. He describes the young researcher's attentiveness and sensitive care for the animals throughout the experiment. He describes the play and socialization periods that were routinely scheduled between researcher and subject. He watched from a distance the very natural human-animal bond develop and express itself. Herzog seems to suggest that the student may have erred in treating the cats "more like pets than research animals," thereby becoming shaky, tense, and withdrawn when the time to kill them ultimately arrived. We are told that the response of colleagues who had become concerned about their friend's emotional reaction was to offer to do the killing for him. They did not apparently attempt to initiate a discussion about his (and their?) conflict. The stand-in offer contains a very important message. It is the message that personal sensitivities or concerns need not or perhaps must not affect the continuation and conduct of research. Sensitivity in this context is seen as a kind of "softness," something like being an unfit athlete. And like an unfit athlete, the proper course is to become trained and hardened through repetition. This training requirement is conditioned and motivated by the value that "the science must go on."

We do not learn whether this second graduate student developed or eschewed an extreme position with respect to the animal research debate as his professional life continued. We are left wondering what this researcher learned from this experience and whether he continued to invest socialization time with experimental animals in his care or whether he adopted a more detached stance by imposing the differential category assignment of "pet" or "experimental subject" to which Herzog alludes.

Research as War

Early in my undergraduate education and continuing throughout graduate school and later professional life, the metaphor of scientific research as a type of war has dominated. This metaphor gives meaning to the previously discussed factors. Tolerating less than optimal physical circumstances; the sense of mission, purpose, and privilege; the urgency; the importance of general ideas over the needs of individuals; the hardening of feeling reactions—all these are made coherent by the metaphor. The typical language of science makes this connection with war perfectly clear. For example, theories are "attacked" by scientists working in the "trenches," the positions of investigators are "shot down" and "ripped to shreds" and "blasted out of the water," laboratory workers are given their "marching orders" and "missions," groups of animals are tested in "squads," and nature is "brought to its knees" and made to "surrender" its secrets. The force of this metaphor should not be underestimated. As Lakoff and Johnson (1980, p. 3) wrote in their influential book on metaphor, "The concepts that govern our thought are not just matters of the intellect. They also govern our everyday functioning, down to the most mundane details."

It may also be the case that in my own field, psychology, the metaphor of science as war has added meaning and significance. Since its formal inception as a laboratory science in the 1870s, there has been a struggle to insist that the discipline justified characterization as a "natural" science and not just a "soft" human science. In other words, as psychologists, we were at war with nature as well as with the scientific establishment.

In a war, traditional morality is set aside. Sacrifice is expected. Comfort is a luxury. All are expected to "suck it up." And so it was in the laboratory. We were expected to work late and put our other lives on hold. That was the model. Students and faculty alike worked into the night without the company of their spouses, girlfriends, and boyfriends. In fact, it was important that the significant others got used to this type of life, for it was likely to be a life-long pattern. Animals were like conscripts in a war, a just war against nature's secrets and society's ignorance. Is it any wonder that concern for animal feelings did not factor in? After all, our own feelings did not count.

The Awakening of Ambivalence

Although my feelings of distress and justificatory uncertainty around the conduct of research with animals got very quiet, the issue never completely disappeared. Certain events and situations tended to keep the questions alive in the background, if not always actively conscious. Notable among them were the paradoxes I saw in the attitudes of one of my graduate school mentors, Harry F. Harlow, best known for his work on maternal deprivation. Although Harlow had been a well-respected research psychologist for 20 years prior to his work on the effects of environmental deprivation and enrichment on monkey behavior, it was not until this work was published that his fame expanded immeasurably. Before this time, Harlow had published extensively on the process of learning and its neural substrates (e.g., Harlow, 1949).

The deprivation work involved limiting and altering the typical pattern of social experiences of newborn rhesus monkeys. The effects of these manipulations were dramatic and devastating. Monkeys reared without mothers and with minimal social experience showed a panoply of bizarre self-directed behaviors such as body rocking, self-clasping, and occasional self-injurious behavior. In social interactions with other monkeys, isolated subjects did not play, explore, or groom. They huddled alone in corners, shunning all social initiations from normally reared or feral-born controls. As adults, these animals continued to show inadequate social behavior in areas including sexual and maternal functioning (see Harlow, Harlow, and Suomi, 1971). Later studies also showed that basic information-processing capabilities were reduced and neural mechanisms concerned with movement were damaged (Beauchamp and Gluck, 1988; Gluck, 1979; Gluck, Harlow, and Schiltz, 1973; Lewis, Gluck, Beauchamp, Keresztury, and Mailman, 1990; Martin, Spicer, Lewis, Gluck, and Cork, 1991).

The sense of importance that came to surround this work helped mute the impact of the difficult daily laboratory routine of seeing hundreds of monkeys, most housed alone in relatively small stainless-steel cages, and participating in various experimental protocols. To get a feeling for the strong momentum that existed for this work, it is important to get a sense of the aura of importance. It was not just a self-created sense of importance, although that was also present. The evidence was everywhere that the profession approved of the work. It was absolutely undeniable that many notable developmental researchers, theorists, and clinicians believed the implications of Harlow's work to be of monumental importance (see Bowlby, 1969). Harlow was constantly traveling throughout the United States and Europe to speak about the research. There certainly must have been critics, but we did not hear about them. In addition, it actually seemed that funding agencies were in competition to provide support for the work.

The funding and support situation in the laboratory was astounding. The laboratory employed a staff that included electrical engineers, carpenters, sheet metal workers, trained observers, consulting surgeons, and animal husbandry experts, all ready and available to make real the research ideas of the faculty and graduate students. Once projects were completed, statistics consultants and editorial specialists were available to help convert the completed project into a published manuscript. Why the work was considered to be so important is a discussion for another time. The point of the present discussion is only that the work of Harlow and his associates received wide critical support from individuals both inside and outside the animal research community. Having said that, I am not supporting the notion that morality is determined by ballot and that an apparent consensus of acceptability obviates the obligation to be involved in ethical analysis. Certainly it does not. What I am saying is that praise, money, and derived prestige are common measures of importance and it was very hard to generate seriously considered questions about the value of the work.

After a while, I did not see the bizarre self-directed behaviors of the monkeys as evidence of suffering. Instead, the rocking, screeching, and occasional self-mutilation of the monkeys had been converted in my mind to topics of study—problems to be understood and then defeated. They were challenges to my intellect and experimental creativity. They did not call to me for compassion, as a yelp from a dog caught underfoot would have. It seemed like a very natural process repeated constantly in the field of biomedical research. A disease process is recognized, explained, and then treated. The difference here that went substantially unrecognized was that the clinical problems we studied were often the products of experiments designed to test broad theoretical questions.

In the face of all of this apparent insensitivity, Harlow was, in fact, one of the few voices in psychology who spoke out against the behavioristic excommunication of feelings and emotions as factors relevant to the discussion and understanding of animal behavior. Harlow taunted the advocates of "muscle twitch" behavioristic psychology by describing monkeys as having rich emotional lives concerned with affectional interrelationships with other animals (Gluck, 1984). Harlow put the paradox right down in front of us. Animals were not Cartesian automatons. Important parts of their lives existed inside them in the form of feelings, even "love." Harlow's early work on learning also supported this cognitivist position. Again, as opposed to describing animal learning as a function of the mechanical strengthening of stimulus-response connections through the action of reinforcement and punishment, he described monkeys as cognitively testing hypotheses about what the correct choices were (Harlow, 1949).

Nonetheless, these complex cognitive and feeling capabilities did not factor into any ethical consideration of experimental justification. Instead, the attribution of these internal states served to further the experimental generalizability of the research, thus strengthening the external validity of the work, which was Harlow's main consideration. He was not particularly interested in monkey behavior per se but only to the extent that the findings were generalizable to the human condition. He openly expressed the idea that a variable shown not to influence nonhuman primate behavior must not, by extension, be relevant to understanding analogous human behavior and ought not be studied (Harlow, Gluck, and Suomi, 1972).

The paradox was there, like a skull at the dinner table. For the most part, it was ignored, suppressed, or rendered invisible. Whenever it did become recognizable, it was easy to attend to the substantial evidence of respectability and importance, of which we have already spoken. But nonetheless it was there, and I could not quite forget it.

By the summer of 1970, I was already trying to plan for graduation. I knew that I did not want to go to another large primate center. I did not like the feel of "big science." I could see the constant pressure on my professor colleagues that had them always looking for a new angle for the upcoming grant deadline. I preferred to seek employment in a small to medium-sized university with a modest department of psychology. If my wish were to materialize, I was going to have to prepare myself to be more self-sufficient. I hoped that it might work out that some department would be interested in establishing, for the first time, a working primate facility. I wanted to be in a position to take such a job. As preparation, I had begun to go on rounds with one of the staff veterinarians, a competent and gentle clinician named Dan Houser. He began to teach me some of the basics of diagnosis and treatment of common primate diseases. He and a laboratory supervisor, Chris Ripp, taught me such things as the delicate technique of giving TB tests (the injection had to be given in a monkey's eyelid, and safely accomplishing this in an unanesthetized monkey was a bit of a trick) and how to treat common diarrheas and bacterial infections that plague monkeys in captivity. Because Chris also ran the nursery, he patiently taught me the basics of infant care.

One Saturday morning, I met Dan at the lab so we could treat a young female stump-tailed macaque who had contracted a shigella dysentery. It was a serious condition that meant the death of the animal if treatment was not engaged early and maintained consistently. We had finished the treatment when one of the caretakers working in another room called for assistance. We went next door to a room that housed a number of female rhesus monkeys and their infants. Each mother and infant pair was housed separately. In one of the cages, we found a mother lying on her side at the bottom of the cage. Her infant clung tightly to her chest with a rather long and stretched-out nipple in her mouth. Once the caretaker had removed the mother from the cage and separated her infant, the problem became starkly evident. The abdomen of the adult monkey was distended and hard. She was suffering from acute gastric dilatation, what is commonly referred to as "bloat." The condition is similar to bloats that affect cattle and horses.

For reasons not well understood, the stomach of a monkey occasionally loses muscle tone, making it difficult or impossible for the monkey to expel the gas generated by digestion. It is not hard to imagine the amount of pain and discomfort that must accompany this process. If the pressure is not relieved, the expanded stomach eventually crushes the liver and heart and kills the monkey. Unfortunately, relief of this condition in those days was hit-or-miss. The veterinarian quickly passed a nasogastric tube into the monkey's stomach and attempted to aspirate the gas and fermenting ingesta with a 50-cc syringe. The procedure failed to relieve the pressure. The monkey's eyes were beginning to lose the luster of life. In a last-ditch effort, a long, large-gauge hypodermic needle was inserted straight into the monkey's stomach on a centerline directly below the sternum. The procedure is something like trying to burst an inflating balloon with a small, thin nail. Unfortunately, in this case the pressure was not relieved.

Meanwhile, a member of the care staff had telephoned the principal investigator, who was a senior-level graduate student, and informed him about the situation. To his credit, he arrived at the lab quickly and was present to observe the vain attempt of the needle procedure. He could no doubt see the condition of the monkey for which he was responsible. At this point, the veterinarian turned to him and asked permission to euthanize the animal. He had, in fact, already drawn a syringe full of a potent barbiturate adequate to painlessly kill the animal. The investigator responded that the animal and her infant were important subjects in an ongoing project. The veterinarian softly countered by saying that he had done all he could do and that it was highly unlikely that the bloat would resolve spontaneously. He continued by pointing out that the animal was in extreme pain. The last word of the discussion was that as long as there was some possibility of resolution, no matter how distant, the animal was not to be killed. He then turned and left the room. The monkey died several hours later.

I cannot honestly say that I was astonished by the decision. This decent, intelligent man represented no more than the typical amount of ego and student arrogance. He was fascinated by monkeys and spent hundreds of hours watching and describing their social behavior. I am sure that he had no wish to see any animal die. He looked for a reason not to kill the monkey and was able to find that reason in the statement that resolution was not 100 percent out of the realm of possibility. He seized on this small probability instead of the reality of the suffering right in front of him. In a way, I even agreed with his decision, at least in the abstract. It was the lack of connection between what was a research decision and the reality of the monkey's current state that did not fit. The anguish of this animal was clearly unrelenting and intense.

I felt sick. The two of us stood there in the treatment room with a beautiful brown, red, and gray monkey who had been wild-caught in some Indian forest or city and brought to this Midwest laboratory some years earlier. She lay there on a polished stainless-steel examination table with the shape of her blue identification chest tattoo distorted by the extreme distension of her stomach. I knew something had gone very wrong. Something very important had been missed. The vet expressed his position by saying that "in research the animal serves the researcher, which in turn obligates us to serve them at times like these." He expressed this sentiment quietly, yet I heard him clearly and understood precisely what he meant. At that moment, I remembered the veterinarian's car parked in the alley behind my childhood home. I remembered the expenditure of anxiety, tears, and the small cash reserves on a very sick, thin, black-and-white dog named Penny, bought from a taxi driver for $2 about two months earlier. We all knew without a formal procedure of balancing costs and benefits that this shivering and vomiting little dog deserved relief, no matter how much we wanted her to stay with us. This conversation with Dan was, in fact, the first time in my undergraduate and graduate animal research experience that I had heard articulated the belief that a human could betray an animal. Until that point, I had come to believe that my obligations in science were only to the experimental structure and the data.

Luckily, I had many more experiences like these, experiences that were reminders of the ambivalence. Most notable among them were the veterinarians that I encountered at other facilities who continued to model the clinical relationship with animals. Under their care and in their hands, experimental subjects were converted to patients whose suffering mattered and required attention and ease. The suffering required relief not only because it interfered or compromised an experiment but also because to create or permit suffering without reason, meaning, or necessity broadens the physical suffering into a moral injury. They insisted that not everything under the guise of research was acceptable and not every research idea was worthy of implementation. In their opinion, some procedures just hurt animals too much. Frankly, all too often they spoke too softly and hesitantly, but the clinical model was consistently there for anyone who had the time and inclination to see it. There were also the students who, unlike me, had the courage to object or question some of the experiments with which I asked them to assist. Robert Frank and Timothy Strongin consistently asked about the shock and food deprivation levels we used in the learning experiments, and they asked about the hypotheses and their importance. These students were not so respectful of the "understood" taboo against discussion of their reactions, and their respect for authority was not quite so clear. There were students who included tributes to the experimental animals in the acknowledgment sections of their theses and dissertations, right next to their spouses, friends, and faculty committee members. At first, I saw these acknowledgments as whimsy, an expression of satire. I came to realize that some of these individuals were quite serious. I came to appreciate that they were trying to express the recognition that they had made use of the life of a sentient entity and that their use required acknowledgment and a public expression of gratitude. There was still little question about the usefulness of the research or a need to weigh and balance the costs of the animals and the value of the research. Rather, it was an acknowledgment of the integrity of another life and an untargeted expression of gratitude. Perhaps the value of the expression was simply in the making of it.

These incidents began to re-create a permeability to the process and necessity of research justification from the point of view of costs to the animal. I began once again to respect the ambivalence and, consequently, a broadened sense of the field of obligations to which I was responsible. I began to feel uncertain about the value of certain experiments again. In doing so, I started to become a moral agent and scientist and not just a designer of experiments.

Getting to the Point

The shape of my laboratory began to change. Certain pieces of equipment were discarded. I began to pile up the shock generators and primate restraining chairs for disposal. I not only wanted them away from my laboratory but also wanted them totally out of use. I was having a difficult time imagining experiments that I could do where their use could be justified. But, most important, I found that I could no longer look with dispassion at the rooms of monkeys living in individual cages. It was as though I was finally appreciating the fact that these animals, capable of expansive intellectual, social, and emotional lives, were living in boxes so that my experiments could be accomplished.

As I watched these animals, all living in individual cages, I saw how many of them attempted to stretch their lives beyond the boundaries of their boxes. I saw how they had learned to see one another by positioning themselves just so, taking advantage of small holes or cracks in the solid side walls of their cages and the reflections from the tiled walls and window glass. I watched as they teased and incited one another from a distance, rattling the cage doors and slamming the metal walls. I watched how they tried to maintain some semblance of social existence in the absence of the opportunity to actually groom, touch, and fight. The monkeys that had been purposely reared without social experience peered out from behind their bizarre postures. They were in two boxes: the metal cage and the behavioral one we had created by limiting their social experience. The experimental questions that I had for these animals paled in the face of what I came to believe was the reality of their misery. Although I was not condemning all animal research, I knew that I could not continue the experiments that I was doing with these animals. The costs and the real benefits were grossly out of balance. These animals had suffered enough.

I quickly discovered that limiting the kinds of experiments utilizing these animals was far more difficult than simply ceasing my own experimentation with them. I was not able to find satisfactory places to send the monkeys. Other primate laboratories were willing to accept the monkeys and continue research with them, which was not acceptable to me. I wanted the experiments with these animals to stop, not just continue out of my sight. The zoos I contacted were interested only in the socialized monkeys and then only if they would continue to breed and produce babies. Unfortunately, these monkeys were well past their reproductive prime.

After many discussions and meditations, I decided that I must euthanize the animals if I wanted the experiments with them to stop. However, I was persuaded that if I was going to take this option, perhaps one more experiment would be permissible. I decided that when I euthanized the animals I would do it in such a way that their brains and spinal cords could be removed and fixed and later studied. A trusted colleague arranged with several respected neurobiologists and pathologists to participate in these in vitro experiments. We all agreed that these were worthwhile experiments, especially because we could not imagine that new primate social isolation experiments would be approved in the improved atmosphere of concern for animal welfare. We believed that these animals were among the last of their kind. These particular animals had given so much that I believed we were obligated to get a last bit of useful information from them in a way that did not increase their burden.

Then in late July, after adequate funding had been received, the team of technicians, scientists, and veterinarians assembled. I tried to explain why I had chosen this course. I reminded them all that I had known each of the monkeys for almost 20 years and that they had all participated in a long stream of experiments. I told them that I believed the burden these animals carried could not justify continued in vivo experiments. The process of euthanasia and tissue collection took a full week to complete. It was very difficult, as it should have been.

I wonder about the word *euthanasia,* which means good death. I do not know if the actual deaths were good. I know we attempted to ensure that a deep plane of anesthesia had been reached prior to the surgery. I checked each monkey for reflexes myself before collection procedures were initiated. I am as certain as I can be that they felt no pain during this, their last experiment. It was truly sad that I thought more about the condition of their death than I had years before thought about the meaning and impact of the experiments to which I had exposed them.

Concluding Remarks

Research with animals is more than an encounter with nature. It is also a moral encounter. Our egoistical needs—our strong desire to describe, explain, predict, and control nature—become pitted against the largely unknown interests of entities unable to engage in shared consent. This circumstance is fraught with the possibility of ill-considered decisions motivated by untempered knowledge seeking and extreme self-interest. Yet, much interferes or lobbies against even this fundamental recognition, a recognition that in my judgement is necessary for the responsible conduct of research with animals. In my research life, the metaphor of research as war has had a particular negative influence in this regard. War is a time when the typical moral order is relaxed and feelings of hesitancy and uncertainty are seen as troublesome obstacles to the "mission." The implied sense of urgency, the feeling of unquestioned importance, and the affirmation of authority that flow from this designation mute the feeling of ambivalence that ought to be present when we use the life of another sentient being for our purposes. There is little room for respect and reflection on ambialence as long as the war metaphor is maintained, either explicitly or implicitly.

Unlike the soldier who was told to treat his rattled nerves after injuring an "enemy" in combat for the first time by "doing it a few more times," scientists must be aware of and resist the pressures to ignore their moral distress. The scientist who no longer feels anything about the subjects in his or her experiments needs to stop and assess what has gone wrong, because something has indeed gone wrong. A researcher who considers the ethical dimensions of a project less worthy of rigorous analysis than the experimental components is only half a scientist.

References

Beauchamp, A., and Gluck, J. P. Associative processes in differentially reared rhesus monkeys: Sensory preconditioning. *Devel. Psychobiol.* 21(4): 355–364, 1988.

Beauchamp, Tom L. The moral standing of animals in medical research. *J. Law Med. Ethics* 20(1–2): 7–16, 1992.

Bekoff, M. Cognitive ethology and the treatment of non-human animals: How matters of mind inform matters of welfare. *Anim. Welfare* 3: 75–96, 1994.

Bowlby, J. *Attachment and Loss.* New York: Basic Books, 1969.

Cheney, D. L., and Seyfarth, R. M. *How Monkeys See the World.* Chicago: University of Chicago Press, 1990.

Cohen, C. The case for the use of animals in biomedical research. *N. Engl. J. Med.* 315(14): 865–870, 1986.

Dennett, D. C. *Brainstorms: Philosophical Essays on Mind and Psychology.* Cambridge, Mass.: Bradford Books, 1978.

Donnelley, S., and K. Nolan, eds. Animals, science and ethics. *Hastings Center Report* (special supplement) May–June, 1990.

Frey, R. Autonomy and the value of animal life. *Monist* 70: 50–63, 1987.

Gluck, J. P. Intellectual consequences of early total social isolation in rhesus monkeys (*Macaca mulatta*). In *Advances in Primatology: Nursery Care of Nonhuman Primates,* G. C. Ruppenthal, ed. New York: Plenum, 1979.

Gluck, J. P. Harry Harlow: Lessons on explanations, ideas and mentorship. *Am. J. Primatol.* 7: 139–146, 1984.

Gluck, J. P., J. Eldridge, and C. D. McIver. Considering our relationships with animals: The message from Zuni pueblo. *Anthrozoos* 6(1): 4–8, 1993.

Gluck, J. P., H. F. Harlow, and K. A. Schiltz. Differential effect of early enrichment and deprivation on learning in the rhesus monkey (*Macaca mulatta*). *J. Comp. Physiol. Psychol.* 84: 598–604, 1973.

Gluck, J. P., and S. Kubacki. Animals in biomedical research. *Ethics Behav.* 1(3): 157–173, 1991.

Griffin, D. *Animal Minds.* Chicago: University of Chicago Press, 1992.

Harlow, H. F. The formation of learning sets. *Psychol. Rev.* 56: 51–65, 1949.

Harlow, H. F., J. P. Gluck, and S. J. Suomi. Generalization of behavioral data between nonhuman and human animals. *Am. Psychol.* 27: 709–716, 1972.

Harlow, H. F., M. K. Harlow, and S. J. Suomi. From thought to therapy: Lessons from a primate laboratory. *Am. Sci.* 59: 538–549, 1971.

Herzog, H. Human morality and animal research. *Am. Scholar* 62(3): 337–349, 1993.

Lakoff, G., and M. Johnson. *Metaphors We Live By.* Chicago: University of Chicago Press, 1980.

Lewis, M., J. P. Gluck, A. J. Beauchamp, M. F. Keresztury, and R. B. Mailman. Long-term effects on early social isolation in *Macaca mulatta:* Changes in dopamine receptor function following apomorphine challenge. *Brain Res.* 513: 67–73, 1990.

Martin, L. J., D. M. Spicer, M. H. Lewis, J. P. Gluck, and L. C. Cork. Social deprivation of infant rhesus monkeys alters chemoarchitecture of the brain: 1. Subcortical regions. *J. Neurosci.* 11(11): 3344–3358, 1991.

Orlans, F. B. *In the Name of Science.* New York: Oxford University Press, 1993.

Plous, S. An attitude survey of animal rights activists. *Psychol. Sci.* 2(3): 194–196, 1991.

Regan, T. *The Case for Animal Rights.* Berkeley: University of California Press, 1983.

Rollin, B. E. *The Unheeded Cry: Animal Consciousness, Animal Pain and Science.* New York: Oxford University Press, 1989.

PART II

CURRENT CONCERNS ON THE RESPONSIBLE CONDUCT OF RESEARCH

Decision Making

4

MELINDA A. NOVAK
MEREDITH J. WEST
KATHRYN A. BAYNE
STEPHEN J. SUOMI

Ethological Research Techniques and Methods

Introduction to Ethology

Ethology is the study of species-typical patterns of behavior. Its aim is to uncover causes, function, developmental patterns, and evolutionary significance of such behaviors (Slater, 1985). Ethological research differs from most biomedical research in a number of ways. A primary difference is that, for ethologists, the animal under study is neither a model nor a surrogate for another species. The animal is a model of how ontogeny and phylogeny operate, but no one species is intrinsically more important than another in describing the myriad ways in which organisms and environments interact over short and long time periods. This approach also emphasizes the uniqueness of different species. Indeed, a fundamental part of ethology is to understand how similar environmental conditions can lead to substantial diversity of behavioral expression both across species and within individuals of the same species (Lott, 1991).

Ethology also differs from biomedical research in the range of species studied. Whereas biomedical disciplines tend to use a few representative species (e.g., domesticated rat or mouse, rhesus monkey, or squirrel monkey) for which there are well-established housing and husbandry practices, ethology includes many other species (e.g., bats, cowbirds, frogs, hamsters, hyenas, sticklebacks, sparrows, tamarins, and voles). Considerably less information is available on what constitutes optimal housing and husbandry for these "exotic" species. Thus, researchers and facilities managers alike may have to attend closely to the reactions of these animals in captivity rather than relying solely on procedures developed for domesticated species in other settings.

An ethologist often strives to see an animal at its species-typical best, behaving in contexts as natural as possible. To do so sometimes means leaving the typical laboratory environment to study animals under free-ranging conditions. An alternative that provides more experimental control is creating seminaturalistic surroundings in the laboratory to stimulate species-typical performance. There is, however, no standard laboratory environment or typical housing arrangement that is synonymous with ethological research. Instead, unique environments are designed by the researcher to elicit and maintain the behavior patterns of interest. Such environments frequently require alterations in husbandry practices (see Gibbons, Wyers, Waters, and Menzel, 1994, for many examples).

For many ethologists, a crucial element of their work is the study of animals in social situations. However, this perspective creates its own concerns about animal well-being. For example, because aggression is a natural part of many species' repertoires, ethologists often must consider how opportunities to observe species-typical social behavior may increase or create aggression. Moreover, ethologists often must consider the welfare of multiple species, only one of whom may be the target of the scientist's interest. The study of predator-prey interactions or optimal foraging, for example, may require the ethologist to weigh the costs and benefits for broader ecological units (Huntingford, 1992).

The very diversity that characterizes ethological research makes it difficult to develop a set of general rules regarding animal care and welfare that apply uniformly to all animals and situations. In fact, procedures that might be appropriate for one species or one setting may not necessarily generalize to other species or other settings for the same species. Thus, our goal is to identify possible welfare issues pertaining to ethological research that may require consideration by investigators, facility managers, and institutional review boards (e.g., institutional animal care and use committees [IACUCs]). The discussion that follows is divided first by setting (natural, seminatural, and laboratory environments) and then within setting by subject matter (passive observation, social manipulations, predation).

The obligation to provide a practical overview of taxonomically quite diverse organisms precludes describing specific experiments with one population of one species or from one laboratory; the information provided has been derived from numerous overviews of animal research practices. Thus, we have provided a minimum of references to specific studies, giving the reader a good place to start. We have also listed specific overviews on aspects of ethological research in the references and provided a source list of professional society guidelines for animal use.

Behavioral Research Settings and Procedures

Natural Environments (Free-Ranging, Islands), Passive Observation

Some ethologists study animals in their natural habitats. Field studies provide information on a species's habitat utilization, foraging strategies, breeding patterns, and social organization and are considered an important adjunct to laboratory studies of similar behavioral phenomena (Altmann, 1980). Such naturalistic studies are often considered relatively benign because they involve passive observation of animals in their natural setting. However, care should be taken to minimize possible harmful effects the observer may have on the population under study and to other populations residing in the study area. These often inadvertent effects include being a vector of disease and interfering with species-typical activities (e.g., serving as a beacon, thereby increasing the risk of predation in prey species or reducing capture rates in predatory species). Use of additional light to study nocturnal animals must also be considered. For example, some wavelengths previously thought not to disturb the behavior of frogs have now been shown to lead to reduced feeding (Buchanan, 1993).

Passive Observation with Provisioning

In some cases, because of the difficulty of observing free-ranging animals, ethologists may resort to provisioning a wild population (i.e., augmenting the food supply in a particular area to attract wild animals). Although this procedure may facilitate observations by bringing animals close to the observer, it also raises important animal welfare concerns (Altmann and Muruthi, 1980). First, there are nutritional considerations. In this regard, the provisioned material should be selected to minimize possible dietary imbalances. Second, provisioning may artificially increase population densities either because of movement of animals into the provisioned area or because of an increase in the breeding rate (Ford and Pitelka, 1984). In some species, increased density may be associated with heightened aggression and, ultimately, lowered reproduction. Third, when the study is over, loss of provisioning may result in high mortality because the environment can no longer support the expanded population. These effects may be controlled partially by considering the frequency and length of the provisioning period as well as the actual distribution of food in terms of the area covered.

Individual Recognition and Sample Collection from Free-Ranging Animals

For some studies of free-ranging animals, it may be necessary to identify individuals and collect biological samples for the purpose of research or health monitoring. Depending on the species, these efforts may require capture, restraint, sedation, and tagging or marking. Animals are typically handled for a brief period and then returned to their habitat.

Capture. The two most common procedures for capturing animals alive involve the use of nets (for birds and bats) and live traps (for small and medium-sized mammals). Permits are usually required for trapping or collecting vertebrates in the United States. These permits are different from fishing and hunting licenses and can usually be obtained from the U.S. Fish and Wildlife Service. Some states may have other regional and local ordinances with additional requirements. If the trapping is to be conducted on privately owned lands, permission of the landowner is also required. Trapping on state or federal lands typically requires some contact with the relevant officials to establish the locations, guidelines, and rules of trapping. When capturing animals, the researcher must consider various welfare issues. Some pertain to the animals themselves; others relate to the person handling the wild animal.

When animals are captured in traps, they are vulnerable to temperature extremes, possible food and water deprivation, and weight loss (Bietz, Whitney, and Anderson, 1977). Thus, traps must be monitored frequently enough to ensure the health and well-being of the trapped animals. Frequency of monitoring depends on the species and both the current and typical weather conditions of the region. It may also depend on the characteristics of nonsubject animals that are likely to enter the traps (e.g., shrews in rodent traps). In addition to the monitoring, provision of materials such as hay or cotton may be needed in the trap under some conditions to offset cold temperatures. Baiting the traps with dry food, fruit, or vegetables can also reduce the chances of possible food or water deprivation.

The use of nets requires relatively constant monitoring inasmuch as animals caught in nets are typically immobilized and cannot be provisioned. Small animals are often removed, handled, and biologically sampled without sedation. For small birds and mammals, the risk of sedation may be greater than the risks associated with brief, expert restraint. Thus, it is important to use restraint techniques that minimize stress and preclude injury. Potential risks to the researchers include injury from bites and scratches and exposure to disease. These risks can be minimized with protective garb, appropriate handling techniques, and immunizations (e.g., prerabies shots when working with some species of small mammals).

Sedation. Some animal species may require sedation for the collection of biological samples (e.g., Sapolsky and Ray, 1989). This technology pertains not only to the handling of larger animals already confined in traps but also to larger animals for which sedation itself becomes the capture technique. Darts and blowguns are frequently used to capture and sedate large animals. Darts or syringes are often used to immobilize animals confined within a trap.

The primary risks in sedating free-ranging animals involve chemical reactions and locomotor disturbances. First and foremost, select the appropriate tranquilizer and delivery route for the species in question. The dosage should be monitored carefully, and antidotes should be available. Because tranquilizers typically affect motor coordination prior to sedation, movements of the subject after darting and before loss of consciousness may require monitoring. This precaution is most important when sedation is the capture technique inasmuch as animals may climb trees or put themselves in other precarious positions before succumbing to the effects of the drug. After sedation, animals should be secured until awakening and released only when their movements will not endanger themselves or others.

Marking. In field research, it may be necessary to make repeated observations of individual subjects. When the subjects are numerous or lack distinctive markings, artificially marking or altering their physical appearance may be the only way to identify them. Techniques vary from applying relatively permanent devices such as leg bands, collars, or ear or fin tags to more temporary alterations involving dye, special paint marks, or radio-tagging devices that become nonfunctional after a period of time. Under some conditions when tags, bands, or collars cannot be used and a permanent mark is required, it may be necessary to alter the physical appearance of an animal (e.g., toe clipping). Marking is usually done in conjunction with trapping and netting or sedation.

Each marking technique has its own costs and benefits, which vary as a function of the species studied. The investigator should select the marking technique that provides adequate individual recognition for the species in question with minimal risk to the subject. The age of the subject to be marked may affect the choice of procedures. For example, although bands and collars provide a relatively safe method for recognizing individual adults, their use with young subjects may be problematic because of the changes in body size and conformation that occur as young animals mature. Some procedures may require surgery, postoperative care, and assessments as to when the animal can be returned to its habitat. For example, although radio-tagging is often accomplished in large animals by containing the transmitter in a collar, in smaller animals, "minimitters" must be inserted under the skin or into the body cavity under appropriate anesthesia. Other procedures may require some treatment. In the case of toe clipping in rodents, the American Society of Mammalogists recommends that animals receive immediate treatment for the injured toes (e.g., antibiotic ointment) and that there be a limitation on the number of toes clipped per individual. Finally, there may be some effect of the marking procedure itself on the individual animal, for example, the animal may behave differently or be perceived differently by species members (Burley, Krantzberg, and Radman, 1982).

Other Manipulations of Field Populations

Although some field studies are primarily descriptive, others are focused on the reactions of animals to environmental changes that are produced through researcher interventions. Such interventions usually take two general forms: (1) alterations of the physical environment and (2) modifications of the social environment.

Alterations of the Physical Environment. The usual purpose of these studies is to examine the factors influencing habitat selection, foraging behavior, breeding biology, and population dynamics. Alterations of the habitat include, but are not limited to, increasing or decreasing cover or shelter (e.g., adding hay) (Taitt and Krebs, 1983), relocating nests or other distinctive features, and adding novel foods or other novel objects (see also the section about provisioning).

Whenever the habitat is altered, there may be changes in breeding rates not only in the subject population but also in other species living in the same area. Furthermore, certain modifications of the habitat may increase the risk of predation. In some cases, these changes may be desirable (e.g., controlling agricultural pests), or they may be the actual focus of the study. In other cases, these risks may interfere with other outcome measures. In species that live in social groups, alteration of the habitat may increase or decrease tension and aggression among group members.

Modification of the Social Environment. The usual purpose of these studies is to understand the communicative process, to reveal features of social organization, or to examine cross-species interactions. The subject population may be exposed to models, to certain sensory cues of animals (e.g., odors or vocalizations), or to living animals. Welfare issues become somewhat more complex as one moves from inanimate presentations to the whole organism.

If the exposure is limited to models, odors, or vocalizations, the risk of dangerous altercations may be reduced (Seyfarth, Cheney, and Marler, 1980). Depending on the species, situation, and presentation schedule, however, there may be some risk of habituation or sensitization to species-relevant stimuli. Furthermore, such presentations may make the subject population more conspicuous to predators. In some cases, these risks are central to the proposed research, whereas in others they may interfere with the planned study.

When the exposure involves a living organism, special techniques may be required for protecting the organism and the subject population from one another (e.g., holding cages). Additional attention should be paid to the stimulus animal's social status if it is a conspecific. Once the exposure is over, there should be some plan regarding the fate of the stimulus animal. Is it to be returned to its original location or incorporated into the subject population, or are there alternatives? The social housing itself should also be considered in terms of its effects on hormonal status, which might affect returning animals to the wild or immediate use in other studies. For example, social isolation of birds and mammals can lead to higher levels of corticosterone and lower levels of testosterone (Dufty and Wingfield, 1990).

Semi-Free-Ranging Situations

A number of species are housed in large groups in outdoor enclosures in zoological parks (e.g., ungulates, rodents, and canids) and in laboratories (e.g., nonhuman primates in outdoor corrals). In some cases, observation of these animals occurs from blinds, catwalks, or other areas that are separated from the animals. In other situations, however, the observers move freely among the animals. When observers and animals can intermingle, there are risks to the health and well-being of both animals and observers. Thus, observers should be knowledgeable about the behavior of the species they are observing. For example, they should be aware of flight distances and not inadvertently corner animals. Before being allowed to observe animals independently, they should receive training from experienced, on-site personnel on how to respond to particular individuals and particular situations and how to protect themselves from danger. Observers need to be screened for the presence of certain diseases that may be highly transmissible to the animals. They should also receive prophylactic inoculations (e.g., prerabies) where relevant.

When animals are housed in large social groups, safe methods must be developed for separating and capturing animals if they should become ill or injured. There should also be a plan for the subsequent return of the animal to its social group. In some primate and avian species, such reintroductions can be problematic, depending on the animal's sex and rank, the length of the time away from the group, and the initial cause of the removal (Bernstein, 1991; West, King, and Eastzer, 1981).

Laboratory Environments

General Considerations

Ethologists frequently strive to incorporate key ecological elements into the laboratory housing situation. In this regard, arboreal species are usually given access to climbing surfaces and structures, scent-marking species are provided with relevant marking surfaces that are not sanitized during every cleaning cycle, and burrowing species are housed under natural covers such as hay. The goal of these efforts may be to stimulate species-typical behavior, to promote physical health and reproductive success, or to introduce controlled variables for experiments (see Gibbons et al., 1994, for examples in rodents, primates, and birds).

Techniques or environments that promote species-typical behavior may seem to interfere with professionally accepted standards of sanitation. However, sanitation objectives need not conflict with "naturalizing" the animal's environment if common sense is applied. For example, hay and other vegetation can be used in conjunction with a suitable schedule for change. Wooden items should be spot cleaned and removed when they show wear. Natural environments need not be unsanitary; however, sometimes restricted cleaning schedules may be appropriate. For example, bat colonies may thrive best when guano is allowed to accumulate in the cage, whereas rigid sanitation schedules may impair their health and well-being (Bayne and Henrickson, 1994). For some rodent species, the transfer of a small amount of soiled bedding to clean cages may actually improve reproductive success (Richmond and Conaway, 1969). Furthermore, scent-marking surfaces should not be routinely cleaned, as this often creates a "strange environment," which for some animals results in excessive scent-marking behavior and physiological stress. This does not mean that every alteration of the laboratory environment that would make the environment more natural is acceptable. Such environments should support both species-typical behavior and well-being.

Wild-Caught Animals as Research Subjects

Sometimes it is desirable to study wild-caught animals in captivity for the purposes of observing behavior under more controlled conditions, evaluating the underlying neural and biochemical mechanisms of behavior, characterizing the species-typical behavior of animals that cannot be readily observed in their natural habitat (e.g., burrowing rodents), or assessing reactions to captivity. The first consideration for wild-caught species pertains to the collection process. Appropriate permits must be obtained for the live capture and subsequent use of animals in captivity. In addition, trapping should be conducted in the manner described previously. The second consideration is the animal's state after capture. When wild-caught animals are brought into the laboratory for research purposes, they typically experience a sudden and dramatic change in their physical and social environment, including exposure to humans and possibly other unfamiliar animals. Thus, special efforts may be required to ensure adequate adjustment to the laboratory environment.

Natural Environments and Sample Collection

Consideration should be given to various demographic characteristics of a species and how they affect the collection and adaptation of this species to the laboratory environment. For example, it may be useful to target animals of a particular age to increase the likelihood of their adaptation to the laboratory environment. Young animals may require significantly more care if they are collected prior to weaning or fledging. Collection of a particular age group may alter the natural population distribution.

A quarantine period is needed for newly arrived animals to protect the health of animals already in the colony, to determine the health status of the incoming animals, and to safeguard the health of personnel. Typically, wild-caught animals have parasites, which, in turn, can carry numerous disease agents. Wild-caught animals can themselves be the vectors of bacterial, viral, and rickettsial agents, many of which are transmissible to humans (e.g., Q fever, hantavirus, salmonella, rabies virus, giardia, Lyme disease). During the quarantine period, the disease status of the animal is determined and treatment is instituted, if necessary. In some cases, animals may be brought into the laboratory to study their pathogenic condition and may require special housing.

The quarantine period also provides time for captured animals to adapt to the laboratory before they are studied in a research project. This conditioning period allows the animal's metabolism to adjust to the new environmental conditions and gives the animal time to recover physiologically, immunologically, and behaviorally from the stress of capture. A second adaptation period may be required if the initial quarantine environment differs substantially from the laboratory housing environment in terms of noise level, room density, ambient lighting, and ventilation rate. The greater the discrepancy between these two environments, the greater the need for two conditioning periods.

It is essential to have an occupational health and safety program. Unprotected exposure to wild-caught animals carrying the viruses and diseases described previously can have potentially negative and possibly fatal consequences for researchers and staff (e.g., caretaker deaths caused by herpes B virus transmitted by macaques and hantavirus transmitted by rodents). Although the essential elements of an occupational health and safety program vary across species, common factors include serum banking, vaccination history, protective clothing, and training of all personnel who come into contact with the animals. Colony management practices can be altered to improve the health and safety of both animals and staff. For example, risk of bites or injury to the handler may be reduced by using transfer boxes rather than relying on direct handling of the animals. This procedure may also reduce animal stress inasmuch as carrying cages create a barrier between the animal and the caretaker. Thus, handling methods that provide the most freedom to the animal without compromising the restraint objective or personnel safety are desirable.

Working with wild-caught species may require redesign of the standard laboratory environment to promote species-typical behavior. This effort may include housing animals in species-typical groups; adding natural materials such as vegetation to the housing environment; altering cages to provide hiding areas, escape routes, and nesting boxes; and varying the diet within the limits of ensuring balanced nutrition. For many nontraditional species, however, the dietary requirements are unknown. Under these conditions, the animals should be observed closely for signs of nutritional deficiencies, in which case novel approaches to diet management may be necessary.

An important concern for those working with wild-caught animals is the final disposition of the animal after experiments are completed. A number of options may be relevant, including euthanasia, placement in another research facility, movement to a "retirement home" (e.g., chimpanzees), or return to the natural habitat. Resolution of this issue depends on a number of practical and ethical concerns. If the animal is to be returned to its native environment, the following should be considered: (1) the likelihood of the animal readjusting to nature with time in captivity as one relevant marker, (2) the specific environment to which it may be returned (i.e., is it the same or similar?), and (3) the possible impact on that environment. Because all three options have costs and benefits depending on the species and the circumstances, it may be necessary to determine the fate of wild-caught animals on a study-by-study basis.

Types of Research

Observation

Some ethological research conducted in the laboratory involves only passive observation of the animals (see Martin and Bateson, 1986, for definitions of categories of behavioral measurement). Usually, the goal is to characterize social organization, breeding biology, or activities, such as parental behavior, under controlled conditions (Wang and Novak, 1994). Laboratory observations may be essential when the species in question is extremely difficult to observe in nature (e.g., nocturnal or burrowing animals). A key element of this research is that animals are studied in laboratory environments that are specially designed to mimic some aspects of the natural environment.

In this research setting, the welfare of the animal may be compromised by aspects of the husbandry program. Conversely, the design of the environment, while promoting species-typical behavior, may interfere with routine husbandry practices. Flexibility and ingenuity are required from both facilities managers and researchers to overcome these problems. For example, alterations in sanitation schedules may be needed when young are being reared. Transfer of some soiled bedding from dirty to clean cages may be crucial for reproductive success in some species.

Observers themselves may be a source of stress for some species because of close proximity. Special blinds, one-way mirrors, or other methods of concealment may be useful for recording the behavior of wary animals. Alternatively, a researcher can provide hiding places for the animals and study their behavior with video equipment. These techniques may work well for highly visual species but may have a limited effect for species whose primary sensory modality is olfaction. When concealment techniques for the observer or animal are not practical or suitable, a period of habituation to the observer may overcome some of the animal's reactions (Caine, 1992).

Environmental Manipulations

This kind of research ranges from studies of habitat and dietary preferences to evaluations of cognitive capabilities. For some of this work, animals are rewarded for appropriate responses with a solid or liquid substance. Here, two considerations may be relevant: how the motivation is achieved and whether the diet is balanced. Some species may have to be deprived of food or liquid to control levels of motivation. Other species may respond for a treat that maintains or supplements their diet. Care should be taken to ensure that the treat does not adversely affect nutrition. When the treat is supplemental to the diet, animals should be monitored for weight gain or for reduced consumption of the standard chow ration, and modifications made accordingly. Various vendors now make "treats" that are nutritionally balanced for nonhuman primates and canines.

Social Manipulations: Exposure to Unfamiliar Animals

Much of the ethological literature is focused on the reactions of animals to members of their own or other species. This research runs the gamut from studies of breeding behavior or group formation to those that examine communication processes. Animals may be exposed to other conspecifics or to specific attributes of those conspecifics, such as their odors or vocalizations. Welfare considerations vary, depending on both the context and the extent of the exposure. For example, when the exposure occurs between two or more unfamiliar animals, care should be taken to minimize the risk of aggression and injury. Creating breeding pairs of some rodent species may require more effort than merely placing the animals in the same cage. To eliminate aggression, males can be placed in a small mesh introduction cage within the home cage of the female and then released several hours later. In some cases, bringing unfamiliar animals together may require the use of introduction cages or other techniques to provide a period of familiarization under controlled conditions (see Nadler, Dahl, Collins, and Gould, 1994, for an example in primates).

Mixed Species Interactions

Occasionally, for scientific purposes, different species may be housed together. One of the more common procedures is to cross-foster young to the parents of a different species in an attempt to unravel genetic and environmental influences on behavior. This approach has been used to study song acquisition in birds, behavioral development in rodents, and patterns of aggression and reconciliation in monkeys. Several cautions should be noted in the cross-fostering paradigm. First, the timing of cross-fostering is generally critical to its success. For some species, fostering must occur within the first day or two of life (e.g., voles; McGuire and Novak, 1987). When the timing is unknown, offspring should be monitored carefully for signs of rejection or neglect. Even when parents care for offspring, continued monitoring for signs of malnourishment may be necessary. Second, there may be significant health risks in housing different species together. Third, cross-fostering can lead to altered species-typical behavior in adulthood (e.g., mating preferences or patterns of parental care; Freeberg, King, and West, 1995; Mason, 1960). Thus, cross-fostered animals may be unsuitable for routine use in breeding colonies because their offspring may differ substantially from the species norm.

Mixed species interactions can be used solely as a husbandry procedure rather than for research purposes. For example, both dogs and cats can serve as companions for some species of nonhuman primates under certain conditions. Different species within the same genus, such as rhesus and cynomolgus monkeys, have been housed together successfully for environmental enrichment. However, caution must be exercised to minimize disease risks to both species. The decision to mix animal species should be based on health status, temperament, and developmental processes. Interactions between two species may change substantially as animals mature.

Another purpose of mixed species interactions is to produce hybrids (Dilger, 1962). In this case, special care should be taken to guard against mixing hybrids with the breeding colony unless that is an explicit objective of the research project. Hybrids and cross-fostered animals should never be returned to the wild.

Separation from Conspecifics During Development

Some research involves separating animals from conspecifics during development. In some cases, the separation is necessary to provide the animal with alternative rearing environments (e.g., rearing nonhuman primates with inanimate surrogates or peers) or with controlled stimulation from conspecifics (e.g., use of playbacks in song acquisition in passerine birds). In other cases, the separation process itself is of interest (e.g., mother-infant separation in nonhuman primates).

When animals are separated from parents through an experimental protocol, the investigator and the animal care staff must assume responsibility for rearing the offspring. Adequate attention must be paid to the temporal provisioning of food, actual food intake, nutrition, warmth, and other biological needs. Consideration must also be given to the possible stress produced by the loss of companions. In this regard, both the timing and the type of separation may be crucial. Offspring that are separated at birth or shortly thereafter may not yet have formed strong social bonds with their parents and peers. In contrast, offspring separated later in development may show acute stress followed by depression in response to separation from conspecifics (e.g., three-month-old rhesus monkey infants separated from their mother). The type of separation also affects the response of the offspring. Separation in which an infant is removed from its social group and placed in a new environment by itself may be considerably different from separation in which a particular conspecific, such as the mother, is removed from the social group and the infant in question remains behind with the other group members. Regardless of the kind of separation, young animals should be monitored closely and evaluated regularly. This discussion pertains to separation during early development and not to removal of juveniles following a natural weaning process, as is the practice of those caring for and maintaining rodent and other breeding colonies (Mineka and Suomi, 1978).

Infanticide

This kind of research examines the response of adults to young offspring and from these findings makes inferences about social organization and patterns of parental care. Historically, this research often entailed injury or death to helpless, defenseless neonates and thus by its very nature was problematic from the perspective of animal welfare. The key issue is the high probability for pain and distress, and as a result, the researcher should determine if the study can be conducted in a way that minimizes risk to the offspring. In some studies, offspring have been placed in a protective barrier (e.g., mesh cage) to reduce the potential for injury from adults. Aggression toward offspring in mesh cages is then used in place of actual killing of offspring. In some species, however, this procedure inhibits the infanticidal response. Alternatively, extensive observation and monitoring are sometimes used to reduce the probability of injury. Adults are observed closely for behavioral signs of imminent attack (e.g., lunges in rodents). When these signs are observed, the adult is then distracted or removed from the testing environment before killing occurs. If these techniques are not feasible, the researcher should use a procedure that minimizes the period of suffering. For example, pups should be removed after a single attack and humanely euthanized.

Predation

Perhaps no ethological research, except infanticide, evokes as much concern as the study of predator-prey relationships. Predation may be studied to examine the behavioral effects of predator presence on prey species (including escape strategies), to examine the predatory behavior itself, or to determine to what extent individuals can acquire predatory behavior for conservation purposes. An understanding of the behavior patterns exhibited by both the prey species and the predator can provide clues to the ecological niche of the animal, the cognitive capacity of the animal (e.g., does it modify its escape or killing behavior based on prior experience?), the sensory capacity of the animal, and its adaptations as a predator or prey.

A major welfare issue in this type of research is the occurrence of pain and injury in animals. It is usually the prey species that is at risk for injury, and it is sometimes possible to protect prey from physical attack with the use of holding cages. However, this procedure is useful only if predators continue to make predatory moves under such conditions. Risk of injury in the prey can sometimes be eliminated by modeling aspects of the predation sequence. For example, prey recognition must occur before the predatory sequence is fully initiated. In many cases, it is not necessary to use live prey for studying this facet of predation, but this strategy cannot be used when movement of the prey is necessary both for recognition and predatory behavior. Although injury is a primary concern for prey, predators can also be harmed by their prey. For example, snakes that kill rodents are also vulnerable to their bites.

Researchers who study predation should consider the well-being of both predator and prey. It is useful to consider limits on the number of times an animal serves as a prey (perhaps once). Furthermore, prey that are wild-caught generally have more experience with predators than laboratory animals and may provide a more accurate portrayal of the true sequence of events. Using a lab mouse rather than a field mouse as prey for a carnivore may not generate a true-to-life rendition of the escape strategies used by the prey and the counterstrategies used by the predator. Similar arguments can be advanced for the predator. As animals forfeit their lives in many predation studies, every attempt should be made to expedite the predation sequence and to minimize the suffering of the prey (Huntingford, 1992).

Summary

Most studies involving animal subjects are not ethological in nature, that is, the animals are typically used as a surrogate for human biological or psychological systems. In many of these studies, however, species-typical behaviors of the animals are used to calibrate the effects of manipulations. Changes in social or cognitive activity, in terms of both quality and quantity of behavior, figure prominently in outcome measures. The guidelines described here indicate the diverse considerations facing ethologists as they discover new behaviors or refine knowledge of other behaviors. The degree of care needed to obtain reliable information about behavior extends to all studies involving animal subjects, and thus many of the general principles described here apply to studies of behavior regardless of setting or rationale.

Guidelines

Academy of Surgical Research. Guidelines for training in surgical research in animals. *J. Invest. Surg.* 2: 263–268, 1989.

Ad Hoc Committee for Animal Care Guidelines. Guidelines for the use of animals in research. *J. Mammal.* 66: 834, 1985.

Ad Hoc Committee on Acceptable Field Methods in Mammalogy. Acceptable field methods in mammalogy. *J. Mammal.* 68 (suppl.): 1–18, 1987.

American Medical Association. Animals in research. *JAMA* 261: 3602–3606, 1989.

American Ornithologists Union, Cooper Ornithological Society, Wilson Ornithological Society. Report of Ad Hoc Committee on the use of wild birds in research. *Auk* 105 (suppl. 1): 1A–41A, 1988.

American Psychological Association. *Guidelines for Ethical Conduct in the Care and Use of Animals.* Washington, D.C.: American Psychological Association, 1985.

American Society of Ichthyologists and Herpetologists and the American Institute of Fisheries Research Biologists. *Guidelines for the Use of Fishes in Field Research,* C. Hubbs, J. G. Nickum, and J. R. Hunter, eds. Lawrence, Kansas: American Society of Ichthyologists and Herpetologists, 1987.

American Society of Ichthyologists and Herpetologists, The Herpetologists League, and Society for the Study of Amphibians and Reptiles. *Guidelines for the Use of Live Amphibians and Reptiles in Field Research.* Lawrence, Kansas: American Society of Ichthyologists and Herpetologists, 1987.

American Veterinary Medical Association. Report of the AVMA Panel on Euthanasia. *J. Am. Vet. Med. Assoc.* 202: 229–249, 1993.

Animal Behaviour Society. Animal care guidelines. *Anim. Behav.* 47: 245–250, 1994.

Animal Welfare Act. 7 U.S.C., 2131–2157. Rev. (1985).

Association of American Medical Colleges; Association of American Universities. Recommendations for governance and management of institutional animal resources. *Lab. Anim.* 15: 34–38, 1986.

Burke, R. E., H. M. Edinger, D. S. Foreman, P. J. Hand, W. Hodos, L. N. Irwin, D. D. Kelly, F. A. King, F. A. Miles, and A. R. Morrison. Guidelines for the use of animals in neuroscience research. *Neurosci. Newslett.* 15: 1–4, 1984.

Canadian Council on Animal Care. *Guide to the Care and Use of Experimental Animals,* vol. 1 (2nd ed.). Ottawa: Canadian Council on Animal Care, 1993.

Council of Europe. *European Convention for the Protection of Vertebrate Animals Used for Experimental or Other Scientific Purposes.* Strasbourg, France: Council of Europe, 1985.

Guidelines for the use of animals in research. *Q. J. Exp. Psychol.* 38B: 111–116, 1986.

International Academy of Animal Welfare Sciences. *Welfare Guidelines for the Reintroduction of Captive-Bred Mammals to the Wild.* Potters Bar, U.K.: Universities Federation for Animal Welfare, 1992.

Larson, J. Services of the Animal Welfare Information Center. *J. Am. Coll. Toxicol.* 7: 262–265, 1988.

National Association for Biomedical Research (NABR). *State Laws Concerning the Use of Animals in Research.* Washington, D.C.: NABR, 1991.

National Research Council. *Education and Training in the Care and Use of Laboratory Animals: A Guide for Developing Institutional Programs.* Washington, D.C.: National Academy Press, 1991.

National Research Council. Institute of Laboratory Animal Resources. *Guide for the Care and Use of Laboratory Animals.* Washington, D.C.: National Academy Press, 1996.

New York Academy of Sciences. *Interdisciplinary Principles and Guidelines for the Use of Animals in Research, Testing, and Education.* New York: New York Academy of Sciences, 1988.

Orlans, F.B., ed. *Field Research Guidelines.* Bethesda, Md.: Scientists Center for Animal Welfare, 1988.

Schaeffer, D. O., K. M. Kleinow, and L. Krulisch, eds. Proceedings from an SCAW/LSU-SVM–sponsored conference, April 8–9, 1991, New Orleans. Greenbelt, Md.: Scientists Center for Animal Welfare, 1992.

Universities Federation for Animal Welfare (UFAW). *Guidelines on the Care of Laboratory Animals and Their Use for Scientific Purposes.* Potters Bar, U.K., 1987.

Working Committee for the Biological Characterization of Laboratory Animals/GV-SOLAS. Guidelines for the specification of animals and husbandry methods when reporting the results of animal experiments. *Lab. Anim.* 19: 106–108, 1985.

References

Altmann, J. *Baboon Mothers and Infants.* Cambridge, Mass.: Harvard University Press, 1980.

Altmann, J., and P. Muruthi. Differences in daily life between semiprovisioned and wild-feeding baboons. *Am. J. Primatol.* 15: 213–221, 1980.

Bayne, K. S., and R. Henrickson. Regulations and guidelines applicable to animals maintained in indoor seminaturalistic facilities. In *Naturalistic Environments in Captivity for Animal Behavior Research,* E. F. Gibbons Jr., E. J. Wyers, E. Waters, and E. W. Menzel Jr., eds. New York: SUNY Press, 1994.

Bernstein, I. S. Social housing of monkeys and apes: Group formation. *Lab. Anim. Sci.* 41: 329–333, 1991.

Bietz, B. F., P. H. Whitney, and P. K. Anderson. Weight loss of *Microtus pennsylvanicus* as a result of trap confinement. *Can. J. Zool.* 55: 426–429, 1977.

Buchanan, B. W. Effects of enhanced lighting on the behaviour of nocturnal frogs. *Anim. Behav.* 45: 893–899, 1993.

Burley, N., G. Krantzberg, and P. Radman. Influence of colour banding on conspecific preferences in zebra finches. *Anim. Behav.* 30: 444–455, 1982.

Caine, N. G. Humans as predators: Observational studies and the risk of pseudohabituation. In *The Inevitable Bond: Examining Scientist-Animal Interactions,* H. Davis and D. Balfour, eds. Cambridge: Cambridge University Press, 1992.

Dilger, W. C. Behavior and genetics. In *Roots of Behavior,* E. L. Bliss, ed. New York: Harper and Row, 1962.

Dufty, A. M., Jr., and J. C. Wingfield. Endocrine responses of captive male brown-headed cowbirds to intrasexual social cues. *Condor* 92: 613–620, 1990.

Ford, F. G., and F. A. Pitelka. Resource limitation in populations of the California vole. *Ecology* 65: 122–135, 1984.

Freeberg, T. M., A. P. King, and M. J. West. Social malleability in cowbirds (*Moluthrus ater artemisiae*): Species and mate recognition in the first 2 years of life. *J. Comp. Psychol.* 109: 357–367, 1995.

Gibbons, E. F., Jr., E. J. Wyers, E. Waters, and E. W. Menzel Jr., eds. *Naturalistic Environments in Captivity for Animal Behavior Research.* New York: SUNY Press, 1994.

Huntingford, F. A. Some ethical issues raised by studies of predation and aggression. In *Ethics in Research in Animal Behaviour: Readings from Animal Behaviour,* M. S. Dawkins and M. Gosling, eds. London: Academic Press, 1992.

Lott, D. F. *Intraspecific Variation in the Social Systems of Wild Vertebrates.* New York: Cambridge University Press, 1991.

Martin, P., and P. P. G. Bateson. *Measuring Behaviour: An Introductory Guide.* Cambridge: Cambridge University Press, 1986.

Mason, W. A. The effects of social restriction on the behavior of rhesus monkeys: I. Free social behavior. *J. Comp. Physiol. Psychol.* 53: 582–589, 1960.

McGuire, B., and M. A. Novak. Effects of cross-fostering on the development of social preferences in the meadow vole (*Microtus pennsylvanicus*). *Behav. Neural Biol.* 47: 167–172, 1987.

Mineka, S., and S. J. Suomi. Social separation in monkeys. *Psychol. Bull.* 85: 1376–1400, 1978.

Nadler, R. D., J. F. Dahl, D. C. Collins, and K. G. Gould. Sexual behavior of chimpanzees (*Pan troglodytes*); Male versus female regulation. *J. Comp. Psychol.* 108: 58–67, 1994.

Richmond, M., and C. H. Conaway. Management, breeding, and reproductive performance of the vole, *Microtus ochrogaster,* in a laboratory colony. *Lab. Anim. Care* 19: 80–87, 1969.

Sapolsky, R. M., and J. C. Ray. Styles of dominance and their endocrine correlates among wild olive baboons. *Am. J. Primatol.* 18: 1–13, 1989.

Seyfarth, R. M., D. L. Cheney, and P. Marler. Monkey responses to three different alarm calls: Evidence of predator classification and semantic communication. *Science* 210: 801–803, 1980.

Slater, P. J. B. *An Introduction to Ethology*. Cambridge: Cambridge University Press, 1985.

Taitt, M. J., and C. J. Krebs. Predation, cover, and food manipulations during a spring decline of *Microtus townsendii*. *J. Anim. Ecol.* 52: 837–848, 1983.

Wang, Z., and M. A. Novak. Alloparental care and the influence of father presence on juvenile prairie voles (*Microtus ochrogaster*). *Anim. Behav.* 47: 281–288, 1994.

West, M. J., A. P. King, and D. H. Eastzer. The cowbird: Reflections on development from an unlikely source. *Am. Sci.* 69: 57–66, 1981.

5

JOHN G. VANDENBERGH

Animal Welfare Issues in Animal Behavior Research

There has been a great deal of thought and emotion expended on the use of animals in research in the past two decades (Paton, 1993). New—and usually more stringent—guidelines and laws have been promulgated, revised, and revised again. It may be time to assess the status of these changes and their impacts on the animals involved, the scientists who study them, and the society that uses the information gathered as a result of their use. A story commonly told in the southern part of the United States may be relevant. During his sermon on Sunday morning, a preacher asks: "How many of you want to go to heaven?"

All respond: "Yes, oh, yes."

"How many want to go right now?" the preacher asks. A long, thoughtful pause was the response from the parishioners.

What appears intuitively good may have some unpredictable consequences, and a "pause," perhaps not long but certainly thoughtful, is in order with regard to the regulation of animals' use in behavioral studies. Such a thoughtful pause is particularly relevant for researchers in animal behavior because most of the regulations in place for the use of laboratory animals focus on animals held in a more typical "biomedical" facility. Animal behaviorists often study animals in the field or in facilities designed to allow the animal to express its natural adaptations, and thus they may have difficulty complying with the guidelines.

Let us begin with four assumptions: First, we are dealing with animal welfare, not animal "rights." Researchers have responsibility for the animals under their control, and I discuss the details of this responsibility, not the rights animals may have. Animal rights is a philosophical and semantic issue and separate from our discussion of animal welfare (Cohen, 1986). Second, concern for and care of animals used in research have varied in the past. Animals have been and occasionally may still be used in a callous, inappropriate manner. However, researchers who use animals today are aware of the need to provide for an animal's welfare and have made significant improvements in animal care in recent years. Third, individuals who study the scientific aspects of animal behavior were drawn to the field because of their interest in animals, and appropriate care for their subjects is very important to the success of their research program. Animals that receive proper care make better research subjects, and when care can be standardized within a species across laboratories, data with less variability are likely to be produced. This rule may be especially true for behavioral studies because behavior is exquisitely sensitive to environmental factors. Fourth, investigators must comply with legal codes and with institutional guidelines. If regulations appear to be inappropriate, attempts should be made to change them through our rule-making process, in the legal system, or within our institutions.

In this chapter, I outline the laws and guidelines, problems, and opportunities that animal behaviorists should consider as they deal with the care and use of vertebrate animals. The focus is on the issues faced by researchers in the United States, but many of the problems and opportunities are similar in other nations.

Laws and Guidelines

The basic federal law that covers the use of laboratory animals in the United States is the Animal Welfare Act, which is promulgated by the U.S. Department of Agriculture (USDA). This law and the other associated regulations and guidelines are described in some detail in the *NABR Issue Update* (NABR, 1993) and the *SCAW Newsletter* (Smeby, 1993). The Animal Welfare Act (AWA) was originally enacted in 1966 and has been amended four times since then, most recently in 1991. Under the AWA, all research facilities that use animals other than laboratory rats and mice must register with the USDA, comply with the specific USDA regulations, be inspected by the USDA, and report annually to verify compliance and to provide the number and species used by type of procedure. The procedures are classified as painless or painful with pain-relief analgesia or anesthesia given or not because of scientific necessity. All registered research facilities must have an institutional animal care and use committee (IACUC). This committee reviews protocols of proposed research on vertebrate animals, gives approval prior to implementing procedures, and inspects facilities twice per year for compliance with the AWA. Animal welfare standards include requirements for handling, housing, and veterinary care. Many of the handling and housing requirements, particularly the requirements that research facilities must have programs in place for exercising dogs and providing for the psychological well-being of nonhuman primates, fall within the purview of animal behaviorists.

The *Guide for the Care and Use of Laboratory Animals* (NRC, 1996), commonly called the *Guide,* is widely accepted by institutions as the primary reference on laboratory animal care and use. The *Guide* was first published in 1963 and has been revised six times to maintain its currency with recently acquired information. It has become the most requested publication from the National Research Council and National Academy of Sciences, with more than 400,000 copies distributed since

1985. The *Guide* was prepared by a panel of laboratory animal veterinarians, biological researchers, and citizens concerned with animal use and ethics. It provides information on the care and use of common laboratory animals and assists investigators in the planning and conduct of experiments using warm-blooded, vertebrate animals. The *Guide* has been adopted by the major professional organizations concerned with the care and use of laboratory animals (the American Association for Accreditation of Laboratory Animal Care [AAALAC] and the American Association for Laboratory Animal Science [AALAS]), by major corporations, and by nine foreign nations as the standard by which animal care and are measured.

In addition to the AWA and the *Guide,* several other policy statements and laws apply to animals in certain circumstances. For example, agencies within the U.S. Public Health Service (USPHS), such as the National Institutes of Health and their grantees, must comply with the *Public Health Service Policy on Humane Care and Use of Laboratory Animals* (USPHS, 1986). This policy statement provides reporting requirements in grant proposals and in reports. It requires programs that receive PHS funding to comply with the AWA and *Guide*. Within the USPHS, the Office for Protection from Research Risks (OPRR) receives reports from IACUCs, handles complaints, attempts to coordinate with USDA, and informs users, the public, and Congress how the welfare of animals is being protected.

Compliance with the Endangered Species Act may be of particular concern to animal behavior researchers. This act prohibits or controls the acquisition of wild or captive-bred animals listed as threatened or endangered. Use of such animals requires appropriate permits. Last, several states and local governments have established local laws or regulations controlling the use of threatened or endangered animals. For more information, see *State Laws Concerning the Use of Animals in Research* (NABR, 1991).

Among this array of regulatory laws, guidelines, and oversight bodies, the most important group to the investigator on a working level is the IACUC. The IACUC is established by the institution and works most closely with the scientists responsible for the care and use of animals. It follows the AWA and the *Guide* in its deliberations and, somewhat like a zoning board, is able to give variances for unusual circumstances. It allows investigators and colony managers to do what they otherwise might be prohibited from doing under strict adherence to guidelines but what might be in the best interest of the animals involved. The role of the IACUC is becoming increasingly important. For example, it has a growing responsibility for scientific merit review as described in the *Public Health Service Policy on Humane Care and Use of Laboratory Animals* (USPHS, 1986). Judging scientific merit is a difficult and potentially contentious issue. Some groups consider that most, if not all, animal research is trivial and duplicative and that peer review of proposals is a "whitewash." Others feel that the value of animal-based research can be high, especially if approved by a peer review body. The IACUCs are going to have to deal with these issues at each institution. A more detailed discussion of scientific merit review is available in Prentice, Crouse, and Mann (1992). The responsibilities of an IACUC are great because its decisions can result in a successful research program at the institution while ensuring that animals are used and cared for in a technically appropriate and humane manner. Overly restrictive application of guidelines can impede research, whereas failure to meet its responsibilities can result in jeopardizing all federal funding at the institution.

Problems

There are several problems related to the existing animal welfare regulations. Primary among them is whether engineering standards or performance standards should be used in assessing the adequacy of an animal's care. Standards based on engineering specifications require compliance with specific measurable components of the animal's environment, whereas standards based on performance provide a standard of the animal's performance (e.g., growth, reproduction) that must be met without specifying exactly how to do it. The "performance" approach has been adopted in the most recent edition of the *Guide* (p. 3)

A good example of the kinds of problems faced in choosing between engineering and performance standards is the issue of ventilation. The standard for ventilation in animal facilities has been 10 to 15 air changes per hour because it appears to provide adequate ventilation for animal facilities; thus, this standard is often regarded as a requirement. The current *Guide* recommends that a minimal ventilation rate should accommodate heat loads based on the biomass and equipment in the room, with the result that either more or less ventilation than the conventional 10 to 15 air changes per hour may be needed.

Meeting performance standards may make it easier to provide for the specific needs of nontraditional species in confinement. Burrowing species such as naked mole rats (*Heterocephalus glaber*), pine voles (*Microtus pinetorum*), hibernators such as 13-lined ground squirrels (*Citellus tridecimliatus*), and even opportunistic burrowers like Norway rats (*Rattus norvegicus*) and house mice (*Mus musculus*) may prefer fewer changes per hour. No data are available on ventilation preferences of commonly used species or health consequences of a specific number of air changes per hour. Some species like snakes may do well with lower levels of ventilation because of their lower rates of metabolism. Other species, perhaps including birds, may require more changes per hour to provide adequate ventilation. Again, no specific information is available. The issue is rarely lack of oxygen but more likely dissipation of waste gases, such as ammonia, that can be toxic at high levels or disagreeable at low levels to humans working in the animal rooms.

Basic engineering standards are a necessary framework in which professional judgment by the scientist and the attending veterinarian can be exercised. Professional judgment is required to deal with the wide array of species and circumstances necessary for the behavioral scientist to conduct research. Ventilation rate is but one of many issues that require this blend of standards and judgment.

Studies of the behavior of animals often have special needs not shared by other subdisciplines of biological or biomedical research. Unusual species are often used because of the comparative nature of the inquiries and because of the ecological relationships under investigation. A summary of the numbers of papers devoted to different taxa in a recent (1994) volume of the journal *Animal Behaviour* (Table 5.1) reveals that 70 percent of the papers published in this representative journal utilized

Table 5.1 Number of full-length papers published in volume 48 of *Animal Behaviour* (1994) by taxa

Taxon	*N*	Percent
Invertebrate	37	30
Vertebrate	87	70
Fish	12	10
Amphibians	3	2
Reptiles	4	3
Birds	39	31
Rodents	13	10
Primates	7	6
Other mammals	9	7

Note: Of the 87 papers published on vertebrates, only 6, or 4 percent, utilized traditional laboratory animals such as mice, rats, and hamsters.

vertebrates, among which only six employed traditional laboratory rodents. Thus, the *Guide* and the regulations related to laboratory animal use and care are unlikely to have specific recommendations applicable to the array of animals used in behavioral research. For example, some species may require special control of the environment not common to traditional laboratory animals, including less frequent cage changing to keep the olfactory environment more constant.

Another example of problems specific to animal behaviorists is the use of equipment in colony rooms. Many studies require long-term observation by videotape, audiotape, or data-collecting equipment, and equipment may have to be left in the room with the animal subjects. Currently, such material left in an animal room can result in a citation from USDA-APHIS or IACUC inspectors.

Another area in which problems are likely to emerge is the applicability of the AWA and the *Guide* to field studies. Field work often entails observation, capture, and handling of species not specifically discussed in the laws and guidelines for animal care and use. Professional societies, such as the Animal Behavior Society, have prepared guidelines (Association for the Study of Animal Behaviour, 1995) that are very helpful to investigators who design studies and to IACUC members who are making judgments on specific procedures. More complete listings of the professional society guidelines are available in Chapter 4 (this volume) and the *Guide*.

When behavioral studies entail procedures that are not specifically covered in the AWA and the *Guide,* the general principles outlined in these documents should be followed. An essential step that can help animal behavior investigators avoid problems is to give the IACUC a clear description of the procedure to be employed and the reasons for using that procedure in the protocol approval process.

Serendipitous findings demonstrate that it is difficult to predict the outcome of research. Many of the major findings in the biological sciences in the past few decades have unexpectedly emerged from exploring the physiological basis of adaptive behavior in animals. For example, early studies by Nottebohm (1985, 1989) and Patton and Nottebohm (1984) on the neural basis of bird song involved surgery, deafening, and lesions to the brain. Few could predict that such procedures would result in a new understanding of regeneration in the central nervous system and stimulate the successful search for sex differences in brain morphology (Nottebohm, 1985) and, as recently shown, brain function (Shaywitz et al., 1995). The rising level of regulation and control may inhibit such research in the future if researchers are forced to predict the outcome or value of their studies.

The costs of complying with stringent engineering regulations can inhibit the progress of research. Many smaller institutions can no longer afford the expenses associated with animal care, and thus investigators and students at such institutions are denied research opportunities. Even at larger, research-oriented institutions, use of animals can be prohibitively expensive, and only well-funded investigators can conduct many types of animal studies. A serious concern is that young investigators who are entering research careers may choose to avoid contentious areas of animal research or use of high-profile species, even though important biological questions remain unanswered in these areas.

Opportunities

Animal behavior studies have a major role to play in improving the welfare of animals used in research. The Animal Behavior Society and its British counterpart, the Association for the Study of Animal Behavior, have formed Animal Care and Ethical Committees, respectively. The Animal Care Committee has recognized that "animal behavior studies are of great importance in increasing our understanding and appreciation of animals" (Association for the Study of Animal Behaviour, 1995, p. 277). Thus, an opportunity exists to gather information important to improve the care of animals and educate the professional and lay public about animals. Studies of the behavior of animals can also contribute to advances in human and animal medicine. For example, the Society for Neuroscience has taken the position that neuroscientists have an obligation to contribute to advances in understanding how the brain works and to further advances in treating and curing disorders of the nervous system through responsible and humane research on animals (Society for Neuroscience, 1991). An excellent example is that provided by the work of Nottebohm and colleagues cited previously. Behavioral research can provide information to assist in making decisions on details of housing, husbandry, and care of animals. Several outlets exist for reports of such studies, including a new journal, the *Journal of Applied Animal Welfare Science*.

A specific need emerged for behavioral research when the 1985 revision of the AWA required exercise plans for dogs and provision for the psychological well-being of nonhuman primates. Without a clear definition of either of these two requirements, investigators and colony managers needed more information. A number of studies have become available, especially on the issue of nonhuman primate psychological well-being (Bayne and McCully, 1989; Crockett et al., 1993; Novak and Suomi, 1991; Reinhardt, 1990). However, more information is still needed on the exercise requirement for dogs. In general, behavioral information is needed on the natural activity patterns of animals and how they interact with their environments to grow and reproduce. This information will permit many animals used in research to have an opportunity to show an array of normal functions in confinement.

An important advance in the effort to ensure the welfare of laboratory animals is the development of reliable behavioral measures to serve as indices of general well-being. Can laboratory environments be "enriched," and can measurements be made of such enrichment attempts? This task is not easy because such measures would have to be specific to species and to the environment experienced by the individual to provide an acceptable measure of well-being.

Two approaches have been taken: (1) enrich the physical environment by structures or enrichment devices or (2) enrich the environment by increasing social contact. An example of the first approach is that of Novak and associates (1995), in which the investigators used a set of behavioral measures to compare laboratory and semi-free-ranging rhesus monkeys (*Macaca mulatta*). They then used this information to naturalize the environment of laboratory-housed monkeys and, finally, tested whether the enriched laboratory-housed monkeys acted more like the semi-free-ranging animals. The result was a qualified yes. Foraging was a big difference between the two groups, as could be expected. When wood shavings were spread on the floor of the laboratory monkeys' cages, differences between indoor and outdoor monkeys were eliminated or reduced in six of nine behavioral categories. However, some changes were judged to be negative. As the authors conclude, "Virtually all forms of environmental enrichment have both costs and benefits that may be distributed unequally across individuals, sexes and species" (p. 42).

Increasing cage size seems intuitively to be a way to improve the housing conditions of animals. Crockett and Bowden (1994) have investigated the issues of cage size and the need for social contact by using laboratory-housed macaque monkeys (*Macaca facicularis* and *M. nemestrina*). Using both long- and short-term behavioral measures and urinary cortisol levels to index physiological stress, they found that cage size was a minor factor in improving their measures of psychological well-being in the monkeys. Much more important were the changes in the routine research and husbandry procedures. The most stressful event for a macaque was to be transferred from its home cage in a familiar room to an unfamiliar cage in an unfamiliar room. Such a transfer caused increased cortisol production, decreased food intake, disrupted sleep, and decreased self-grooming during the first day after transfer. This information indicates that keeping transfers of monkeys to a minimum can significantly improve overall psychological well-being much more than providing a larger cage.

Changing the social environment has also been examined. Whary and associates (1993) evaluated the use of group-housed laboratory rabbits (*Oryctolagus cuniculus*) versus the almost universally used isolation cage. Single- or group-housed rabbits did not significantly differ on a number of physiological and immunological measurements. Evaluation of behavior indicated that the rabbits preferred small social groups, selected specific microhabitats within the enclosure, and displayed a wider array of behaviors when housed in groups. From their data, Whary and colleagues concluded that group housing enriched the environment of the rabbits without affecting variables important to the investigators who were using the rabbits in physiological or immunological studies. However, young, nonbreeding female rabbits were used in this study so potential problems with aggression were avoided. Group housing of adult males would quite likely result in severe aggression.

Most primates are also highly social animals, and many facilities have developed strategies for housing nonhuman primates in social groups (see review by Reinhardt, 1990). However, social grouping is not without risk to the animal. In a systematic study of the extent of wounding in macaques housed individually, in pairs, or in groups, Bayne and associates (1995) found a number of effects. Pair housing resulted in the highest level of wounding, 33 percent, followed by 13 percent among group-housed monkeys, and 5 percent for the individually housed monkeys. Further, females acquired wounds more frequently than did males (adult males were apparently not housed together), and seasonal differences in wounding occurred with a peak in spring.

Many variables can influence wounding in caged monkeys. Crockett and Bowden (1994) found that gender plays an important role. Fifteen female-female pairs were compatible when introduced into a cage and displayed behaviors indicating improved psychological well-being. In contrast, only one-third of the paired males showed similar compatibility. Ranheim and Reinhardt (1989) have shown that choosing compatible individuals through a trial-and-error process can reduce the incidence of wounding. This study and others like it show that a general rule, such as "monkeys should be paired to enhance psychological welfare," may not be appropriate. Pairing males is counterproductive, and even females have to be matched for compatibility. Further, some rules, such as "bigger cages are better," may not be the case as both Crockett and Bowden (1994) and Bayne and McCully (1989) have shown. Psychological welfare can be enhanced more effectively by reducing the number of cage transfers. An appropriate give-and-take between the rule makers, rule enforcers, and animal users can result in an effective animal care program that takes advantage of the behavioral biology of the species and does not impede research.

Conclusions

Although animal behaviorists have contributed significantly to the improvement of animal care, more work is needed. Information is needed on several levels. First, information is needed on the natural behavior of species kept in captivity so that laboratory housing can be designed to take advantage of the animals' natural adaptations. Second, more information is needed on responses to specific management issues, such as inedible inanimate enrichment (toys) for caged animals, perches, and appropriate alternative caging. The techniques that have been successfully applied to understanding the mechanisms of behavior and its consequences for the survival and reproductive success of species must now be applied to the more structured environment of the laboratory. Also, and perhaps most important, a framework or theoretical basis for behavioral studies focused on improvement of animal husbandry is needed. Continued collection of data in empirical studies will provide the grist for the theorists' mill.

ACKNOWLEDGMENTS I thank T. E Hamm, S. P. McGinnis, and K. S. Wekesa for helpful comments on earlier drafts of the manuscript. This work was made possible through the support of the North Carolina Agricultural Research Service, Project 06262.

References

Association for the Study of Animal Behaviour. Guidelines for the treatment of animals in behavioral research and teaching. *Anim. Behav.* 49: 277–282, 1995.

Bayne, K., M. Haines, S. Dexter, D. Woodman, and C. Evans. Nonhuman primate wounding prevalence: A retrospective analysis. *Lab Anim.* 24: 40–44, 1995.

Bayne, K. A. L., and C. McCully. The effect of cage size on the behavior of individually housed rhesus monkeys. *Lab Anim.* 18: 25–28, 1989.

Cohen, C. The case for the use of animals in biomedical research. *N. Engl. J. Med.* 315: 865–870, 1986.

Crockett, C. M., and D. M. Bowden. Challenging conventional wisdom for housing monkeys. *Lab Anim.* 23: 29–33, 1994.

Crockett, C. M., C. L. Bowers, G. P. Sackett, and D. M. Bowden. Urinary cortisol responses of long-tailed macaques to five cage sizes, tethering, sedation and room change. *Am. J. Primatol.* 30: 55–74, 1993.

NABR (National Association for Biomedical Research). *State Laws Concerning the Use of Animals in Research,* ed. 3. Washington, D.C.: National Association for Biomedical Research, 1991.

NABR. Regulation of biomedical research using animals. *Issue Update,* pp. 1–4, 1993.

Nottebohm, F. "Hope for a new neurology." Neuronal replacement in adulthood. *Ann. N. Y. Acad. Sci.* 457: 143–161, 1985.

Nottebohm, F. From bird song to neurogenesis. *Sci. Amer.* 260: 74–79, 1989.

Novak, M. A., A. Ruff, H. Monroe, K. Parks, C. Price, P. O'Neill, and S. J. Suomi. Using a standard to evaluate the effects of environmental enrichment. *Lab Anim.* 24: 37–42, 1995.

Novak, M. A., and S. J. Suomi. Social interaction in nonhuman primates: An underlying theme for primate research. *Lab. Anim. Care* 41: 308–314, 1991.

NRC (National Research Council). *Guide for the Care and Use of Laboratory Animals.* Washington, D.C.: National Academy Press, 1996.

Paton, W. *Man and Mouse,* ed. 2. New York: Oxford University Press, 1993.

Patton, J. A., and F. Nottebohm. Neurons generated in the adult brain are recruited into functional circuits. *Science* 225: 1046–1048, 1984.

Prentice, E. D., D. A. Crouse, and M. D. Mann. Scientific merit review: Role of the IACUC. *ILAR News* 34: 15–19, 1992.

Ranheim, S., and V. Reinhardt. Compatible rhesus monkeys provide long-term stimulation for each other. *Lab. Primate Newslett.* 28: 1–2, 1989.

Reinhardt, V. Social enrichment for laboratory primate: A critical review. *Lab. Primate Newslett.* 29: 7–11, 1990.

Shaywitz, B. A., S. E. Shaywitz, K. R. Pugh, R. T. Constable, P. Skudlarski, R. K. Fulbright, R. A. Bronen, J. M. Fletcher, D. P. Shankwieler, L. Katz, and J. C. Gore. Sex differences in the functional organization of the brain for language. *Nature* 373: 607–608, 1995.

Smeby, R. R. Summary of new animal welfare regulations. *SCAW Newslett.* 15(1–3): 7, 8, and 4, respectively, 1993.

Society for Neuroscience. *Policies on the Use of Animals and Humans in Neuroscience Research.* Washington, D.C.: Society for Neuroscience, 1991.

USPHS. *Public Health Service Policy on Humane Care and Use of Laboratory Animals.* Bethesda, Md.: Office for Protection from Research Risks, National Institutes of Health, 1986.

Whary, M., R. Peper, G. Borkowski, W. Lawrence, and F. Ferguson. The effects of group housing on the research use of the laboratory rabbit. *Lab. Anim.* 27: 330–341, 1993.

GORDON M. BURGHARDT

Snake Stories

From the Additive Model to Ethology's Fifth Aim

Two-Headed Problems

At the National Museum in London is a startling jade sculpture of an Aztec god: a snake with a head at each end of its body. Native Americans of the Northwest Coast still frequently portray the Sisiutl, a dangerous two-headed serpent that guarded the homes of spirit creatures. Surely these are mythological beings; real snakes have only one head.

But myths are never isolated from reality; myths reflect aspects of the world as perceived by others. Accordingly, some herpetologists might interpret these deities as based on real snakes, but ones in which the tail has evolved features to mimic the snake's head. There actually are snakes with this feature, such as the rubber boa (*Charina bottae*) of North America and the red sand boa (*Eryx johnii*) of India. Indeed, the fake head may even be used in a display to divert a predator to the wrong end of the snake's body and thus allow the real head and most of the body to escape (Greene, 1973, 1988). Perhaps the living animals that inspired these gods did not really have two heads, and one just appeared that way. The fake head was an illusion, albeit a potentially useful one if the fake head deflected attacks by naive people and nonhuman predators (Whitaker, 1978). But as scientists we now can and should go beyond such primitive folk psychology (see chapter 7).

Many two-headed dichotomous issues in science are also treated as myths. A common response has been to act like the herpetologist and dismiss one head as a mere deception. No such issue has been more discomforting to behavioral scientists, conceptually and politically, than those surrounding competing claims for dichotomies such as nature versus nurture, genetic versus environmental, or innate versus acquired. Similarly controversial are the resurgence of individual versus group selection (Wilson and Sober, 1994) and selfish versus altruistic behavior. A somewhat related dichotomy is that between biological and cultural evolution (Boyd and Richerson, 1985). Often a proponent of one side of such dichotomies claims that some new insights or data resolve a dichotomy or reveal it as a nonissue or mere illusion. But before many years pass, the same theme comes to the fore again in a barely modified guise. The phenomena involved just will not go away. Dogmatic assertions, appeals to authority, word play, redefining terms, clever analogies, and "definitive" experiments may "scotch the snake but not kill it." And this is as it should be; the existence of dichotomies that enrage passions can actually advance science by inspiring innovative research.

Today, the field of animal behavior must deal with the revival of another two-headed dichotomy, this one captured in distinctions such as body versus mind, monism versus dualism, materialism versus mentalism, instinct versus mind, objective versus subjective, sentient versus nonsentient, and conscious versus unconscious. The overlapping issues captured in these terms are not just philosophical preoccupations beyond the realm of science; they reflect real phenomena that must be addressed in the study of human and nonhuman behavior. To varying degrees, our explicit or implicit stance on these issues has serious ramifications for what we study in animal behavior, how we treat the animals we study, and the kinds of interpretations we draw from our results. This book is based on the recognition that the landscape and moral ecology of animal research is shifting rapidly.

Two-Headed Snakes

The vast majority of snakes are monocephalic. But to insist that all snakes *must* have only one real head because of some inevitable law of nature is just not true, regardless of the logic. Two-headed snakes not only exist in nature but also are more frequent than any other dicephalic vertebrate (Cunningham, 1937). Therefore, the deception explanation for the two-headed mythological snake is not the only hypothesis grounded in the biological world. The true explanation may not be that one head is real and the other illusion but that both are different, equally real characteristics of the same organism. Bicephalic snakes may also be the source of ancient legends of multiheaded hydras.

For many years, my colleagues and I have been privileged to study a male two-headed black rat snake (*Elaphe obsoleta*). Both heads, although on the same end of the body (Figure 6.1), are complete and have full sensory and behavioral capabilities (e.g., tongue flick, capture and ingest prey). From the first, we noted that both heads might attack and attempt to swallow the same mouse (Burghardt, 1991). In fact, "battles" would go on for an hour or more if we did not intervene, although one head would usually prevail in any given encounter. The two heads reminded me, over 20 years ago, of the perennial but often suppressed conflict between physiological and mentalistic approaches to understanding complex animal behavior (Burghardt, 1978). The snake was thus named *IM,* the left head being *Instinct* and the right head *Mind.* As part of the same body, Instinct and Mind could not be viewed as totally independent, nor could one head be viewed as more "real" or important than the other.

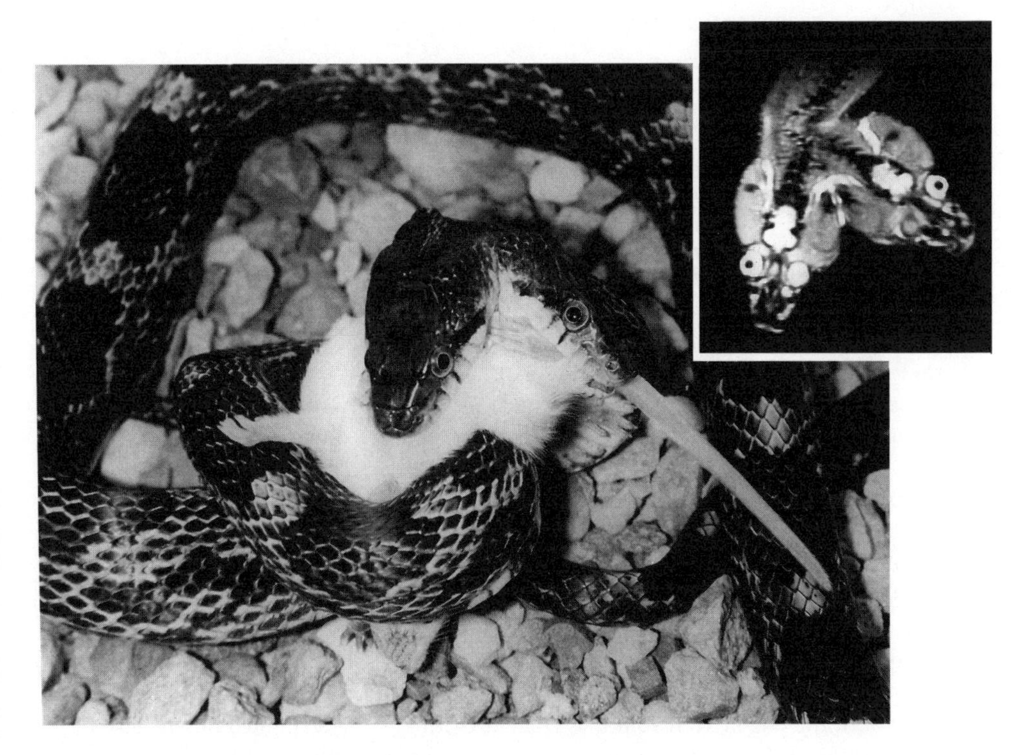

Figure 6.1 IM, the two-headed black snake.

The Merging of Issues in Animal Behavior Research and Animal Care

For more than half a century, the aims of ethology have stayed close to the four classic problem areas summarized by Niko Tinbergen (1963). These areas can be listed as causation (mechanism, control), function (adaptedness, survival value), evolution (phylogeny), and development (ontogeny). Tinbergen's position was that all four approaches were necessary for understanding any behavior pattern; they all function as determinants and consequences across different time frames. Although this comprehensive view was often ignored as scientists embraced one aim or the other as being the most important, the wisdom of Tinbergen's integrative perspective has been repeatedly recognized. However, even as reformulated more recently (Dewsbury, 1992), it is evident that some new fields, including animal welfare, applied ethology, conservation biology, cognitive neuroscience, psychoneuroimmunology, and computer simulation do not rest easily within the Tinbergen agenda.

Many of these new hybrid fields can be recast into the classical four aims and actually enrich their analysis. They force us to pose and answer our questions in altered applied and interdisciplinary contexts. But the dichotomies reflected by IM's name can be accommodated only partially by the traditional four aims. A radical shift seems needed in our basic conception of the nature of knowledge we can obtain from other animals.

The Need for a Fifth Aim

Two major movements have changed the playing field since Tinbergen (1951) asserted, in line with conventional scientific thought, that science could not deal with the subjective lives of animals or the world as experienced by animals.

The first movement was the recognition that behaviorism was unable to convincingly deal with phenomena associated with core issues that psychology should study rather than explain away (e.g., Baars, 1994). These issues included cognitive processes (e.g., ideas, memories, expectations, language, images, representations), social processes (e.g., attachment, conflict, prejudice, communication), and emotional and motivational phenomena (e.g., hunger, fear, anger, jealousy, addictions, psychological disorders). Although the early rise of cognitivism in psychology was based on a rather mechanistic approach that distanced itself from the subjective aspect of cognitive events (Burghardt, 1985), this is no longer the case. The respectability of consciousness, especially, rose in fields from philosophy (e.g., Dennett, 1991) to physics, biology, and psychology (e.g., Baars, 1994). A veritable flood of books on consciousness and mentality in humans, nonhuman animals, and even computers and robots has appeared in the last five years. More than 20 years ago, at the start of the cognitive revolution, Donald Griffin (1976) exploited the malaise of behaviorism in the study of animals by advocating a cognitive ethology. Soon, issues of awareness, thinking, and consciousness in animals were in the forefront of discussion more than they had been in the preceding 60 years (e.g., Griffin, 1982). Since then, issues of animal ability and mentality have been increasingly popular, as recent issues of the *Journal of Comparative Psychology* attest (see also chapter 9). For example, the application of psychological methods, such as Piagetian mental development stages, to animals has been quite extensive. Although there is much controversy about the value of various approaches to the animal mind, researchers increasingly recognize that important phenomena mirrored in animal mentality must be addressed (Mitchell, Thompson, and Miles, 1997).

The second movement was the demand that animals be treated more humanely in the home, school, laboratory, factory, and farm. There have even been demands for suspension of almost all uses of animals by people (discussed in chapter 11). Singer's *Animal Liberation* (1975) reflected and channeled the extension of liberal, egalitarian concerns from all areas of human social life to the treatment of other animals (Midgley, 1983). This perspective soon meshed with the willingness of ethologists, influenced by the first movement, to discuss phenomena previously tainted by anthropomorphism (Rollin, 1989).

Over the last 20 years, ethological and psychological studies have confirmed the surprising complexity of much animal behavior and the apparent similarity of many physiological and communicative processes in nonhuman and human animals. Many students of animal behavior began a bolder questioning of research methods and housing practices in captive animal research. Thus, when selected scientific studies of animals were challenged as egregious examples of alleged mistreatment and callousness, respected scientists were often on both sides of the issue. The wide publicity given stories such as the baboon head injury project energized a receptive vocal constituency that fought, politically, for changes (chapter 11). Although calls for the abolition of all animal research have failed, mechanisms were put in place in government, industry, and universities to check abuses and to minimize pain and suffering in animal care and experimental protocols. Now, U.S. laws mandate that the psychological well-being of primates be addressed. Many European countries have even more restrictive regulations on the captive housing of a wide range of species. Studying animals with the concepts and methods of cultural anthropology is also being advocated (Noske, 1989, 1993).

Animal welfare and treatment issues involving physical and psychological suffering are not separable from the quality of life and experience a captive animal is granted (Bekoff, 1994; Donnelley and Nolan, 1990; Rollin, 1989). Yet, concern for how animals experience the world has been dismissed by many as, at best, unresolvable personal opinion; at worst, sentimental nonsense; and, in any event, beyond the capabilities of natural science (see Burghardt, 1994; Rakover, 1990, 1995). For example, an advocate of environmental enrichment for captive animals recently wrote that the "subjective state of what the animal feels . . . is still not measurable" (Stricklin, 1995, p. 27). Nonetheless, implementing the mandate to enhance "psychological well-being" for captive primates has fostered extensive discussion (e.g., Box, 1991; Gibbons et al., 1994; Mench and Krulisch, 1990; Novak and Petto, 1991; Novak and Suomi, 1988), for it requires that we actually assess and measure well-being. The real debate seems to be whether we should adopt the strategy used in maintaining our family car or seek inspiration from how we provide for our children.

The issues go far beyond animal suffering, pain, and stress, the typical concerns of animal welfarists (Rollin, 1989), although this literature contains many sophisticated and refined methods (Rowan, 1995). More broadly, how do other beings relate to and experience their world? Jacob von Uexküll (1909/1985) wrote two prophetic sentences: "Our anthropocentric way of looking at things must retreat further and further, and the standpoint of the animal must be the only decisive one" (p. 223) and "Animals construct nature for themselves according to their special needs" (p. 234).

To formally recognize the importance and timeliness of assessing the way other species perceive and process events, I proposed adding to Tinbergen's four aims of ethology a fifth aim: the analysis of private experience (Burghardt, 1995, 1997). It is not only the study of processes labeled cognition, awareness, or consciousness but also study of mood, perception, feelings, and motivation. A level of intimacy with a species is required. Complete knowledge of the privately lived experience or inner world of another organism, including another human being, is impossible (Baars, 1994). Yet partial understanding is possible, and obtaining it is essential for scientific reasons as well as for moral ones, namely to fulfill our obligation to treat well our nonhuman research subjects. As a pioneering ethologist with more than 55 years in the field recently wrote, "I think that research on the subjective worlds of animals holds the epistemological clue to the most fundamental problem of understanding animals" (Kortlandt, 1993, p. 140). The determinants and consequences of private experiences need study from the perspectives of the other four ethological aims as well. Conversely, issues of private experience must be at least considered in analyses of the standard four aims.

The Need to Replace the Additive Model

Any merit to the preceding comments is for nought if we have no means to gain information about the lives that animals experience. Veterinarians relying on observations of physical health have objective measures to assess the value of their interventions, but can ethologists and psychologists amass comparable support for their claims? If they cannot, then physical health measures of well-being win by default, for they are based on "objective" evidence.

If we hold there is no way to obtain scientific data on psychological well-being and quality-of-life experience for animals, then how do we improve them? The typical solution is to adopt the *additive model* of animal welfare and to be tyrannized by it. I have presented this model at length in relation to zoos and aquariums (Burghardt, 1996). Briefly, in this approach, common in institutional settings from farms to zoos and laboratories, the first and primary welfare goal is to deal with the animal's physical condition; encouraging natural behavior and well-being is secondary. Only when risks to the animal's health and physical appearance are minimal is it permissible, but not really necessary, to focus on behavior and psychology. In other words, if the animal appears physically healthy and is in clean surroundings in cages of some standard size, our obligations are met. At this most elementary level, the house veterinarian makes the determination of whether adequate facilities are provided. Often animal welfare and rights proponents accept this approach, arguing for minimal cage sizes and fixed maintenance procedures on the basis of scant objective information on how the animals have commerce with the world. This "structural" standards approach to animal welfare is often contrasted to one based on "performance" standards. In the performance standard approach, the ways in which animals use and respond to their housing conditions are the benchmarks, not the often arbitrary physical standards of caging and physical health found in regulations (de Waal, 1989; Novak and Petto, 1991). Performance standards are clearly a move in the right direction, but only if performance really means more than physical appearance and enhances the captive lives animals lead.

Adopting the additive model also implies acceptance of a body-mind dualism that posits a mental world of experiences separable and different from the physical, bodily state of the animal. Although the model accepts the fact that good physical health can produce a contented animal, it ignores the reverse, that natural behavior and a stimulating environment can facilitate physical health.

Additional limitations of the additive model come from studies with humans, which indicate that a focus on physical health and even material resources is insufficient for their well-being. In people, subjective well-being (happiness) is related to health and access to resources (e.g., money) only when they are lacking. Conversely, good health and resources do not in any way assure a fulfilling life. What does facilitate happiness in people is a feeling of being in control, as well as immersion in behavior the individual feels is worthwhile (Myers and Diener, 1995). For animals, too, giving them control over what they do and when and providing them opportunities to express diverse behavior may be essential (Markowitz, 1982; Markowitz and Gavazzi, 1995) for their well-being. A stress-free, bland existence is not the main end in life for any animal.

If this logic holds, then improving the quality of life experienced by animals does not mean reducing all risks and stress to the maximum extent possible. The additive model, however, assumes that risks to physical condition always outweigh other considerations; for example, researchers are prohibited from providing animals with natural substrates for "sanitary" reasons. Carnivorous species are often denied the benefits of naturalistic feeding on live prey by fears of parasites, public criticism, possible suffering of the prey species, or the resulting "messy" cage or by lack of funds, adequate facilities, and staff time. Many reptile keepers feed snakes dead prey because of convenience and fear of injury to their charges. All these are legitimate concerns, but the idea that animals may actually benefit from performing a natural and important, often complex, part of their behavioral repertoire seems rarely considered. Should a small, albeit real, risk to a few individuals always trump a definite advantage for all animals? I once nursed back to health an orphaned black bear cub in my home. She had a splint on her leg, for the bear (Amelia Bearhart) had fallen out of a tree in her enclosure and had broken her leg. Should we now deprive all captive black bears of climbing opportunities? Quality of life is not necessarily enhanced by minimizing every conceivable risk, although the nature of the risks permitted should be related to the genre of the research and by objective data about these risks.

Ironically, attempts to improve the quality of life for captive animals by calling for environmental "enrichment" actually encourages the persistence of the additive model. It makes a concern for lives experienced by animals something extra—nice if you can provide it but not really necessary—such as a rich dessert following a nutritious meal. These attitudes may be more true of research facilities than of public animal exhibit institutions, although the problem exists there, too (Burghardt, 1996).

An alternative to the additive model of animal welfare is the *controlled deprivation model* (Burghardt, 1996). This approach, based on Hediger's (1950) pioneering insights, necessitates identifying and providing the resources animals need—nutritional, climatic, physical, spatial, structural, and social—based on their evolutionary adaptations at all levels. The focus is on creating environments that speak to the entire animal.

Hediger (1950) emphasized that the secret to understanding animals in captivity was to identify and closely analyze those crucial differences from the wild state that affected their behavior. Thus we need to learn what the species does in the wild and the relevant environmental stimuli underlying both perception and response. In short, Hediger's message was that we need to know what the animal is being deprived of in captivity and the consequences of each kind of deprivation. All animals in captivity are deprived of some conditions that may affect their behavior in differing ways. Our obligation is to provide animals with those aspects of nature that are critical for the most important aspects of their lives, behaviorally and psychologically, and not limited to physical survival and health. The animal and its environment must be viewed as an integrated system. Of course, animals in the wild can exist in widely differing conditions of food, environmental structure, predator risk, and almost all other factors that are manipulated in captivity. Many species are highly adaptable in both field and captivity, a fact that can confuse us as to the nature of healthy populations as compared with healthy individuals.

Nonetheless, we delude ourselves if we think that captive-reared animals, even highly domesticated ones, are not severely affected by being deprived in many of the important senses articulated by Hediger: amount and quality of space, food, social structure, climatic variables, and social bonding processes. We must attempt to selectively provide animals an environment that minimizes adverse effects—from the animal's perspective, not ours. Thus, while it might be advantageous for wild animals to have few or no interactions with people, captive animals may benefit from a personal relationship with their keepers (Davis and Balfour, 1992). Behavioral training can reduce the need for anesthesia or physical restraint in veterinary and research procedures such as taking blood or giving injections (Reinhardt, 1992).

In many ways, the change in emphasis this conception demands is already recognized, as in the value of short-term stressors on primate well-being (Novak and Suomi, 1988). For example, Moodie and Chamove (1990) found that exposing tamarins to a hawk model for brief periods was beneficial in that the variety of expressed natural behaviors increased.

Physiological measures involving the immune system are also useful in assessing the effects of social and environmental interventions, which provide data in packets that the biomedical and veterinary communities can accept. Coe and Scheffler (1989) found that rehousing squirrel monkeys and disrupting social groups caused psychological disturbance that could be measured by an immune response, specifically a decrease in blood leukocytes. Carlstead, Brown, and Seidensticker (1993) found that leopard cats that were moved to barren cages showed elevated urinary cortisol, which declined when concealment areas were provided. These studies, among many on stress effects in animals, really gain their explanatory power because of the similarities of these changes in bodily fluids with comparable findings in people. Would we be comfortable trapped in bare rooms or with obnoxious people? This empathic anthropomorphism is just one way in which we can enhance our understanding of the private experience of other species.

Entering the Circle

Von Uexküll and Kriszat (1934) presented a most useful diagram called the *Funktionkreiz* or functional circle (Figure 6.2). It formed the basis of von Uexküll's concepts of the *Umwelt* (surrounding world) and *Innenwelt* (inner world). Stimulus cues from objects are recognized by the receptors, passed to the central nervous system (CNS) perceptual and motor systems, and result in effector action that acts on the object. This process alters the subsequent cues received by the organism and how they are evaluated by the CNS. Animals have several functional circles in which different perceptual cues and behavioral responses are linked together. Major circles are those concerned with food, predators, mates, rivals, and offspring.

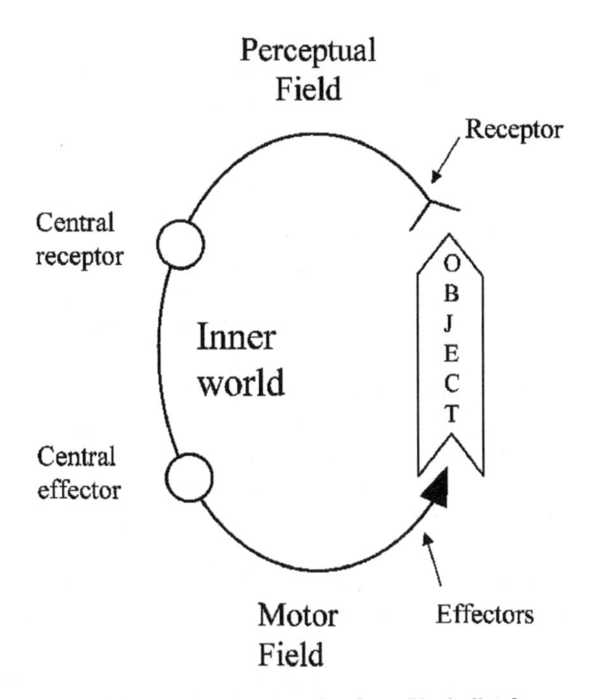

Figure 6.2 The functional cycle of von Uexküll (after von Uexküll, 1934).

The "inner world of the subject" aspect of von Uexkull's functional circle was largely ignored in mainline ethology and comparative psychology (see Kortlandt, 1993). Von Uexküll's view (1909/1985), echoed by Hediger (1950), was that the animal's perceptual world and inner or phenomenal world are the most important aspects of life to any animal (see also Bekoff, 1994; Duncan and Petherick, 1991; Kortlandt, 1993). In psychology, Gibson (1966) independently pioneered an influential theoretical framework that also focuses on the environment as perceived by the organism, not on the stimuli measured by the engineer.

We should not be derailed by our inability, at least currently, to actually inhabit another organism's inner world. We need to distinguish between first-person and third-person descriptions of experiences and mental events in other animals. Just because we cannot directly experience a dolphin's subjective events or how the world appears to a cat does not mean that we can gain no information about such phenomena that has scientific support and predictive value. I cannot directly experience my daughters' inner events or those of their pet rats. Yet, I possess considerable knowledge about what will enhance or diminish the qualities of their lives.

The most effective parents, teachers, farmers, animal trainers, zookeepers, and researchers have an implicit understanding of this knowledge and, most important, know how to apply it. The method for going beyond a subjective or empathic anthropomorphism (subjective analogical inference) is to utilize *critical anthropomorphism,* which entails applying our knowledge of ourselves and empathy in critical concert with scientific knowledge of natural behavior, sense organs, nervous system, social organization, ecology and preferred environments, physiology, evolution, and individual history of members of the other species (Burghardt, 1991; Morton, Burghardt, and Smith, 1990).

Serpents as a Test Case for Critical Anthropomorphism

As a struggling working ethologist trying to understand my favorite research subjects, snakes, it is only fair to consider whether these ideas have had any impact on my own research program. Frankly, most of my reptile behavior colleagues feel a bit leery of applying critical anthropomorphism to reptiles. Unlike my psychologist colleagues, their suspicions are not derived from a muted but still lingering ideological behavioristic stance or from an allegiance to a rigid canon of parsimony. It is because reptiles do not seem to have the kind of emotional life, intelligence, and behavioral flexibility that seem so characteristic of many mammals (e.g., Sjolander, 1995).

The apparent alien nature of reptiles is actually an advantage for testing the role of critical anthropomorphism, however, because a crude projective or empathic anthropomorphism (Lockwood, 1985) is more easily recognized as fallacious or at least not very useful. Furthermore, the sensory, neural, and motoric differences between humans and snakes are substantial. Even a reciprocal personal bond with a specific human caregiver is hard to document in snakes (Bowers and Burghardt, 1992). But effects of stress, leading to chronic and acute alteration and debilitation of behavior have been documented in reptiles (Greenberg, 1992; Guillette, Cree, and Rooney, 1995; Warwick, 1995).

Here, I will briefly discuss three different snake research problems with which I have been involved. The first example derives from subjective analogical inference, the second raises questions of phenomenal experience, and the third shows how new information changes a cognitive interpretation of behavior to a more motivational one.

Deception

Hognose snakes (*Heterodon*) have an elaborate antipredator display that, after bluff attacks and a vigorously simulated dying scene, culminates in a most realistic feigned death (Burghardt, 1991; Greene, 1988). Early attempts to explain the behavior acknowledged that the behavior might deter predation, but that the cause was an epileptic-like fit or a stress-induced adrenalin surge. The exquisite details of this behavior sequence made the uncontrolled "fit" or "shock" explanations highly suspect, to say the least. Taking a cue from prior work on tonic immobility in birds (e.g., Gallup, Nash, and Ellison, 1971) and lizards (Henning, Dunlop, and Gallup, 1976), Harry Greene and I hypothesized that, if we were the snake playing dead in front of a predator, we would monitor the predator closely and leave as soon as the predator left the area or, failing that, leave when the predator was not gazing directly at us. A series of experiments with hatchling snakes who were feigning death in different conditions supported, as shown by faster recovery times, both predictions. This result was the first evidence that the hognose snake was actually alert and aware of its surroundings during the time it was on its back, immobile, not breathing, with mouth open and tongue extruded (Burghardt and Greene, 1988).

These experiments showing that the eyes are themselves a visual cue in the psychology of predator-prey relations are not unique, even among reptiles. More recent experiments have documented the attention that garter snakes (Herzog and Bern, 1992) and iguanas (Burger, Gochfelt, and Murray, 1992) pay to the eyes of simulated predators. Perhaps the characterization of human eyes as "windows of the soul" has more than metaphorical meaning, reflecting an ancient evolutionary history.

The Color of "Yuck"

The second example involves garter snakes (*Thamnophis*), diurnal predators on small prey. Chemical cues are primarily involved in recognizing prey and eliciting bites, but visual cues are readily used in orienting toward, following, and capturing prey (Ford and Burghardt, 1993). Lithium chloride (LiCl) causes illness in many vertebrate species. In snakes, it is indicated by lethargy and vomiting of recently ingested food. Such illness can lead to conditioned food aversions (Burghardt, Wilcoxon, and Czaplicki, 1973). The nature of these experiences of "illness" and food aversions in animals is often compared with human experience (the sauce Béarnaise phenomenon) (Seligman and Hager, 1972).

In a recent experiment, we fed plains garter snakes (*Thamnophis radix*) prey using control or conspicuously colored and patterned forceps (Terrick, Mumme, and Burghardt, 1995). The pattern used, black and yellow stripes, is a common aposematic (warning) pattern found in distasteful prey (Guilford, 1990). When feeding was followed by LiCl injection, all snakes developed an aversion to the prey experienced (earthworm), but not to control prey (fish). However, those snakes fed with the forceps with the aposematic markings showed a stronger aversion than those fed the same prey on a plain background. Unexpectedly, future recognition of the unpalatable prey depended on its chemosensory characteristics rather than its visual badge; that is, snakes who became sick after eating the prey offered on the warning-patterned forceps showed a much greater aversion to it when later offered safe presentations on unpatterned forceps. Snakes made sick after receiving prey on unpatterned forceps showed a much weaker aversion when later offered prey on the patterned forceps. Thus, in garter snakes, warning or aposematic patterns enhance the learning of unpalatable prey but are not later used to identify such prey! Perhaps the effect was mediated by attentional or motivational mechanisms. In any event, this result was a unique twist on the typical findings in aposematic prey research and, apparently, the first demonstration of a cross-modal effect in responses to aposematic prey (contra claim of Rowe and Guilford, 1996).

Animals that avoid warning-colored prey are, reasonably enough, assumed to be able to see the color of the prey—a safe assumption with insect-eating birds and lizards (Guilford, 1990; Sexton, 1960). Color is, however, a subjective, personal experience concept. Colors are not "out there"; they are translations of stimuli processed through eye and brain (see Landesman, 1989). However, to date, no one has demonstrated wavelength discrimination, the prerequisite for color vision, in any snake. But this fact should not be surprising because snakes have rarely been used in traditional sensory discrimination tasks of any type. The retinas of garter snakes do contain three morphologically different cones, which strongly suggests the presence of color vision (Underwood, 1970). Furthermore, the adaptive value of learning to avoid brightly colored distasteful prey was clear.

Shortly after this paper (Terrick et al., 1995) appeared, a leading researcher on comparative vision sent me a report from his laboratory on electrophysiological measurements of wavelength sensitivity of the three cone types in two species of *Thamnophis.* Surprisingly, all three cones have the exact same maximal sensitivity (Jacobs et al., 1992), and thus it seems that the snakes could not have been responding to wavelength, let alone color, in our experiment. Perhaps the dark and light bands resulting from the black and yellow pattern is the effective aposematic cue in a monochromatic animal (Terrick et al., 1995). More research is necessary, but this example shows the dangers of an anthropocentric inference.

Thus, a seemingly conventional study of how animals learn to discriminate safe from dangerous food brings in issues of the private experience (illness, aversion, color and pattern vision) of snakes. No matter how objective we make our language, the issues of what the world means to the animal studied cannot be avoided.

Cooperating Heads

Could IM, the two-headed snake, also be more than a metaphor for some perplexing quandaries that face the study of animal behavior and ethology? My third example involves IM and the role of new information in the interpretation of the private experiences that may be occurring. For the first five years we had IM, we recorded in detail all instances when the snake was offered food; both heads were free to capture and swallow live mice. Many times, both heads struck the prey and attempted to swallow the mouse simultaneously. We found no diminution of fighting over the five years. We thought we might see some cooperation because the food was going to nourish the same body and, indeed, going to the same stomach (supported by radiographic evidence; Burghardt, 1991).

Over the five years, we recorded data suggesting that some "cooperation" between the heads might, nevertheless, be taking place. When Bruce Batts, Brian Bock, and I compared the total mass of prey eaten by each head over the five years, they were virtually identical, as were profitabilities when handling time was factored in (Burghardt, 1991). In interpreting these results, I wanted to resolve this conflict between two measures of cooperation: serious fighting and resource partitioning. To do so, I invoked my twin daughters, then only several years old. The observations made then hold today. While the twins can be remarkably considerate, helpful, and generous with one another, they can also get into the most vicious altercations over what seem to their parents to be minor issues. I thought it reasonable to postulate that cooperation as rational as we might expect has not had a chance to evolve, in either two-headed snakes or human identical twins, commensurate with their convergent genetic interests. Identical twins share more interests and preferences than do fraternal twins. This similarity may both intensify conflicts over the same items and push them to solutions and ways of avoiding conflicts in their more deliberate moments.

New information on IM and the application of critical anthropomorphism now allow a more satisfying and economical interpretation of IM's behavior. We had always assumed that he had a common digestive tract; radiographs taken in 1977 when IM was very young (and small) indicated a common stomach (Burghardt, 1991). Before publishing more recent experiments (Andreadis and Burghardt, 1993) in which the interpretation was critically dependent on the number of stomachs, we carried out extensive video radiology experiments with the help of Robert Toal at the University of Tennessee Veterinary Hospital. We were able to observe and videotape the swallowing process in IM and discovered that IM has two complete digestive systems, at least into the colon. Now, suddenly, the virtually equal prey ingestion by the two heads could be interpreted in motivational terms: The apparent cooperation in equalizing the amount eaten by each head could be an outcome modulated by the fluctuating satiety of each head-stomach unit.

Communication as Window

All these snake examples involve predator-prey interactions. For many species, social behavior may be a richer context for the study of private experience. Griffin (1976, 1982) has advocated using communication as a window to the animal mind, including human-animal communication. To accomplish this poses great difficulties, although much intriguing and controversial information is available (Cavalieri and Singer, 1993; Griffin, 1992).

I have taken two of von Uexküll's functional circles and combined them to create a model for communication across individuals (Figure 6.3). Note that the same stimulus is perceived and acted on by both organisms. Insofar as the behavior and responses of the two individuals merge, to that extent the two "inner worlds" converge (intersubjectivity). Is there any experimental way of showing the usefulness of this model, especially in a nonprimate?

Consider the study by Lubinski and Thompson (1993) on communication based on private states. In brief, in this experimental model that utilizes operant conditioning, a pigeon was trained to communicate its subjective feelings to another pigeon. One pigeon was given injections with different psychoactive drugs, which presumably led to different bodily states, and was trained to signal with key peck responses which drug it had received. By pressing a key, another pigeon could ask the injected pigeon to press a key associated with the private (subjective) state engendered by the specific drug with which it had been injected. The drugged bird accurately replied on the basis of its awareness of its bodily state because it had no way of knowing which drug it had been given.

This example is in some respects a meshing of two of von Uexküll's functional circles. It would be more convincing if both pigeons had direct knowledge of the drug states and could ask whether they had the same or similar states or symptoms.

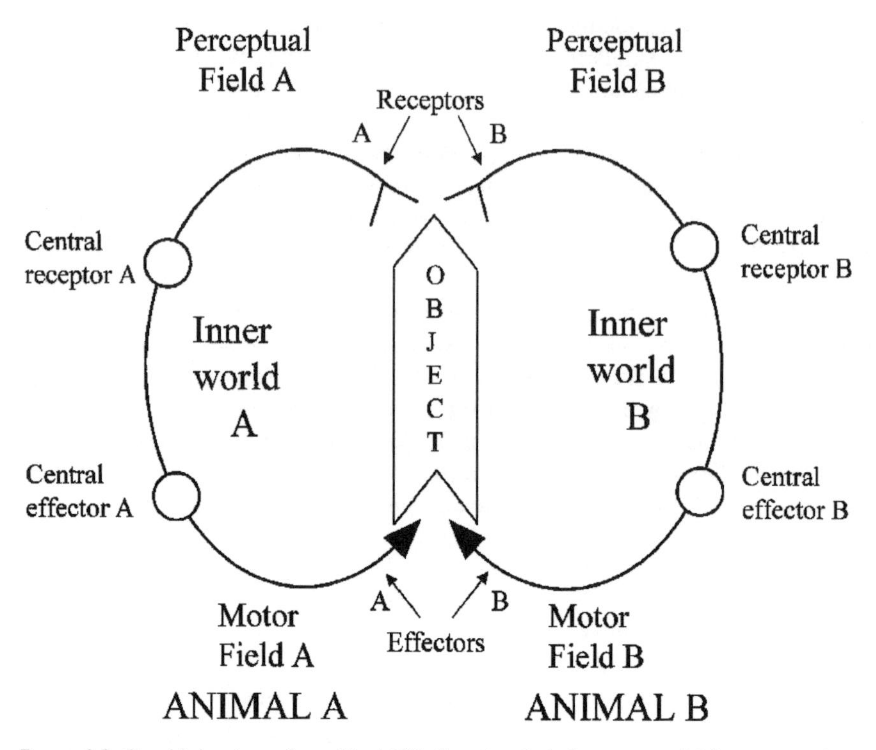

Figure 6.3 Combining two of von Uexküll's functional circles as a model for communication as a process that involves private experience.

Nonetheless, there may be many ways to tap into these systems, and more such work should soon appear. It will be very controversial, as indicated by the spirited commentaries following the Lubinski and Thompson (1993) paper.

New Directions

Currently, there are many means to gain behavioral information about how other organisms process information, make decisions, and respond emotionally. Especially exciting is the potential for new noninvasive methods of entering the circle. Brain imaging of mental processes and abilities is growing rapidly in sophistication, especially with respect to people (Posner and Raichle, 1994). Correlational studies of behavior and brain activity within and across species should become commonplace. Subjective experiences and cognitive events are being correlated with electrophysiological measures in humans (e.g., Paller, Kutas, and McIsaac, 1995), and extensions to animals are sure to proliferate. Virtual reality methods are also available to address *Umwelt* issues. In one pioneering experiment, tethered flies were placed in simulators surrounded by a visual space that changed in response to the animal's "flight" as measured by recording from its flight muscles (Dickenson and Lighton, 1995). In this way, the visual cues that the fly uses to orient in space can be studied.

Although the initial experimental applications of brain imaging and virtual reality have been to study problems other than private experience, their usefulness for the von Uexkull and Griffin agendas seems clear. Convergence in brain processes, even more than in behavior, will provide powerful evidence for the nature of the private experiences of other species. As such information is gained, distinctions among the physical, physiological, behavioral, and psychological will recede and the tyranny of the additive model will end. Animals will become important testing grounds for theories on subjective experiences, emotions, and decision processes. Such research will have important consequences for enhancing the welfare of all captive animals (Gibbons et al., 1994). However, note that this "neural analogical inference" is meshed with both "subjective analogical inference" and subsidiary information of many kinds (Burghardt, 1985).

In the three snake examples, I used largely behavioral methods of reaching conclusions as to the nature of experiences snakes have and why it matters. A view of snakes as having an inner world worth investigating is leading to experiments to elucidate environmental and self-awareness (Chiszar et al., 1995). Such experiments often need imagination more than advanced technology. Currently, we are also carrying out magnetic resonance imaging (MRI) studies of garter snakes (see Figure 6.1) in the hope that we can trace the activity of prey chemical responses in the CNS and further understand the nature of their private experience as well as their physiology. All such behavioral and physiological studies in animals raise questions concerning the similarities and differences of the phenomenological processes involved across species. The additive model, which attends primarily to the physical well-being of captive animals, must be replaced by a controlled deprivation model that can view a social or behavioral deficiency as potentially more salient than a vitamin deficiency.

The nature of perceptual, cognitive, motivational, and emotional processes must differ among species. The extent and nature of the differences are not yet clear. Nonetheless, researchers and textbook writers almost invariably bring in terms such as *hunger* and *thirst* to discuss mechanisms that control intake of food and water (e.g., Grier and Burk, 1992), as well as the modulation of *fear, anxiety, depression,* and *rage* in human and nonhuman species. Often the words used are claimed to be merely convenient shorthand terms with no experiential referent and certainly no claim of similarity across species. But I still wonder about the vitality of those old dichotomies that have so often been put down and then resurrected. With the new tools available, we can now seriously consider the degree to which we can characterize and use predictively these presumptive indicators of the nature of how other species experience the world.

ACKNOWLEDGMENTS I thank Paul Andreadis, Marc Bekoff, Dan Cunningham, Donald Griffin, Harold Herzog Jr., Don Owings, R. Mark Waters, and reviewers for helpful suggestions. The research reported here was supported by several National Science Foundation research grants awarded since 1975.

References

Andreadis, P. T., and G. M. Burghardt. Feeding behavior and an oropharyngeal component of satiety in a two-headed snake. *Physiol. Behav.* 54: 649–658, 1993.

Baars, B. J. A thoroughly empirical approach to consciousness. *Psyche* 1(6): 1–18, 1994.

Bekoff, M. Cognitive ethology and the treatment on non-human animals: How matters of mind inform matters of welfare. *Anim. Welfare* 3: 75–96, 1994.

Bowers, B. B., and G. M. Burghardt. The scientist and the snake. In *The Inevitable Bond,* H. Davis and D. Balfour, eds. Cambridge: Cambridge University Press, 1992.

Box, H. O., ed. *Primate Responses to Environmental Change.* London: Chapman and Hall, 1991.

Boyd, R., and P. J. Richerson. *Culture and the Evolutionary Process.* Chicago: University of Chicago Press, 1985.

Burger, J., M. Gochfeld, and B. G. Murray Jr. Risk discrimination of eye contact and directness of approach in black iguanas (*Ctenosaura similis*). *J. Comp. Psychol.* 106: 97–101, 1992.

Burghardt, G. M. Closing the circle: The ethology of mind. *Behav. Brain Sci.* 1: 562–563, 1978.

Burghardt, G. M. Animal awareness: Current perceptions and historical perspective. *Am. Psychol.* 40: 905–919, 1985.

Burghardt, G. M. Cognitive ethology and critical anthropomorphism: A snake with two heads and hognose snakes that play dead. In *Cognitive Ethology: The Minds of Other Animals,* C. A. Ristau, ed. Hillsdale, N.J.: Erlbaum, 1991.

Burghardt, G. M. Evolution and the analysis of private experience. Psycoloquy 94.5.73. metapsychology.6.burghardt, 1994.

Burghardt, G. M. Brain imaging, ethology, and the nonhuman mind. *Behav. Brain Sci.* 18: 339–340, 1995.

Burghardt, G. M. Environmental enrichment or controlled deprivation? In *The Well-Being of Animals in Zoo and Aquarium Sponsored Research,* G. M. Burghardt, J. T. Bielitzki, J. R. Boyce, and D. O. Schaeffer, eds. Greenbelt, Md.: Scientists Center for Animal Welfare, 1996.

Burghardt, G. M. Amending Tinbergen: A fifth aim for ethology. In *Anthropomorphism, Anecdotes, and Animals,* R. W. Mitchell, N. S. Thompson, and H. L. Miles, eds. Albany, N.Y.: SUNY Press, 1997.

Burghardt, G. M., and H. W. Greene. Predator simulation and duration of death feigning in neonate hognose snakes. *Anim. Behav.* 36: 1842–1844, 1988.

Burghardt, G. M., H. C. Wilcoxon, and J. A. Czaplicki. Conditioning in garter snakes: Aversion to palatable prey induced by delayed illness. *Anim. Learn. Behav.* 1: 317–320, 1973.

Carlstead, K., J. L. Brown, and J. Seidensticker. Behavioral and adrenocortical responses to environmental changes in leopard cats (*Felis bengalensis*). *Zoo Biol.* 12: 321–331, 1993.

Cavalieri, P., and P. Singer, eds. *The Great Ape Project.* New York: St. Martin's Press, 1993.

Chiszar, D., W. T. Tomlinson, H. M. Smith, J. B. Murphy, and C. W. Radcliffe. Behavioural consequences of husbandry manipulations: Indicators of arousal, quiescence and environmental awareness. In *Health and Welfare of Captive Reptiles,* C. Warwick, F. L. Frye, and J. B. Murphy, eds. London: Chapman and Hall, 1995.

Coe, C. L., and J. Scheffler. Utility of immune measures for evaluating psychological well-being in nonhuman primates. *Zoo Biol.* 1(Suppl.): 89–99, 1989.

Cunningham, B. *Axial Bifurcation in Snakes.* Durham, N.C.: Duke University Press, 1937.

Davis, H., and D. Balfour, eds. *The Inevitable Bond: Examining Scientist-Animal Interactions.* Cambridge: Cambridge University Press, 1992.

De Waal, F. B. M. The myth of a simple relation between space and aggression in captive primates. *Zoo Biol.* 1(Suppl.): 141–148, 1989.

Dennett, D. *Consciousness Explained.* Boston: Little Brown, 1991.

Dewsbury, D. A. On the problems studied in ethology, comparative psychology, and animal behavior. *Ethology* 92: 89–107, 1992.

Dickenson, M. H., and J. R. B. Lighton. Muscle efficiency and elastic storage in the flight motor of *Drosophila. Science* 268: 87–90, 1995.

Donnelley S., and K. Nolan, eds. *Animals, Science, and Ethics. Hastings Center Report* 20(3): 1–32, 1990.

Duncan, I. J. H., and J. C. Petherick. The implications of cognitive processes for animal welfare. *J. Anim. Sci.* 69: 5017–5022, 1991.

Ford, N. B., and G. M. Burghardt. Perceptual mechanisms and the behavioral ecology of snakes. In *Snakes: Ecology and Behavior,* R. A. Seigel and J. T. Collins, eds. New York: McGraw-Hill, 1993.

Gallup, G. G. Jr., R. F. Nash, and A. L. Ellison Jr. Tonic immobility as a reaction to predation: Artificial eyes as a fear stimulus for chickens. *Psychonomic Sci.* 3: 79–80, 1971.

Gibbons, E. F. Jr., E. J. Wyers, E. Waters, and E. W. Menzel Jr., eds. *Naturalistic Environments in Captivity for Animal Behavior Research.* Albany, N.Y.: SUNY Press, 1994.

Gibson, J. J. *The Senses Considered as Perceptual Systems.* Boston: Houghton Mifflin, 1966.

Greenberg, N. The saurian psyche revisited: Lizards in research. In *The Care and Use of Amphibians, Reptiles and Fish in Research,* D. O. Schaefer, K. M. Kleinow, and L. Krulisch, eds. Bethesda, Md.: Scientists Center for Animal Welfare, 1992.

Greene, H. W. Defensive tail display by snakes and amphisbaenians. *J. Herpetol.* 7: 143–161, 1973.

Greene, H. W. Antipredator mechanisms in reptiles. *Biol. Reptilia* 16: 1–152, 1988.

Grier, J. W., and T. Burk. *Biology of Animal Behavior,* ed. 2. St. Louis: Mosby, 1992.

Griffin, D. *The Question of Animal Awareness: Evolutionary Continuity of Mental Experience.* New York: Rockefeller University Press, 1976.

Griffin, D, ed. *Animal Mind—Human Mind.* Berlin: Springer-Verlag, 1982.

Griffin, D. *Animal Minds.* Chicago: University of Chicago Press, 1992.

Guilford, T. The evolution of aposematicism. In *Insect Defenses: Adaptive Mechanisms and Strategies of Prey and Predators,* D. L. Evans and J. O. Schmidt, eds. Albany, N.Y.: SUNY Press, 1990.

Guillette, L. J. Jr., A. Cree, and A. A. Rooney. Biology of stress: Interactions with reproduction, immunology and intermediary metabolism. In *Health and Welfare of Captive Reptiles,* C. Warwick, F. L. Frye, and J. B. Murphy, eds. London: Chapman and Hall, 1995.

Hediger, H. *Wild Animals in Captivity.* London: Butterworth, 1950.

Henning, C. W., W. P. Dunlop, and G. G. Gallup Jr. Effects of defensive distance and opportunity to escape on tonic immobility in *Anolis carolinensis. Psychol. Rec.* 26: 313–320, 1976.

Herzog, H. A. Jr., and C. Bern. Do garter snakes strike at the eyes of predators? *Anim. Behav.* 44: 771–773, 1992.

Jacobs, G. H., J. A. Fenwick, M. A. Crognale, and F. F. Deegan III. The all-cone retina of the garter snake: spectral mechanisms and photopigment. *J. Comp. Physiol. A* 170: 701–707, 1992.

Kortlandt, A. Spirits dressed in furs? In *The Great Ape Project,* P. Cavalieri and P. Singer, eds. New York: St. Martin's Press. 1993.

Landesman, C. *Color and Consciousness.* Philadelphia: Temple University Press, 1989.

Lockwood, R. Anthropomorphism is not a four letter word. In *Advances in Animal Welfare*

Science 1985/86, M. W. Fox and L. D. Mickley, eds. Washington, D.C.: Humane Society of the United States, 1985.

Lubinski, D., and T. Thompson. Species and individual differences in communication of private states. *Behav. Brain Sci.* 16: 627–680, 1993.

Markowitz, H. *Behavioral Enrichment in the Zoo.* New York: Van Nostrand Reinhold, 1982.

Markowitz, H., and A. Gavazzi. Eleven principles for improving the quality of captive animal life. *Lab Anim.* 24(4): 30–33, 1995.

Mench, J. A., and L. Krulisch, eds. *Well-Being of Nonhuman Primates in Research.* Bethesda, Md.: Scientists Center for Animal Welfare, 1990.

Midgley, M. *Animals and Why They Matter.* Athens: University of Georgia Press, 1983.

Mitchell, R. W., N. S. Thompson, and H. L. Miles, eds. *Anthropomorphism, Anecdotes, and Animals.* Albany, N.Y.: SUNY Press, 1997.

Moodie, E. L., and A. S. Chamove. Brief threatening events beneficial for captive tamarins? *Zoo Biol.* 9: 275–286, 1990.

Morton, D. B., G. M. Burghardt, and J. A. Smith. Critical anthropomorphism, animal suffering, and the ecological context. *Hastings Center Report* 20(3): 9–16, 1990.

Myers, D. G., and E. Diener. Who is happy? *Psychol. Sci.* 6: 10–19, 1995.

Noske, B. *Humans and Other Animals: Beyond the Boundaries of Anthropology.* London: Pluto Press, 1989.

Noske, B. Great apes as anthropological subjects—deconstructing anthropocentrism. In *The Great Ape Project,* P. Cavalieri and P. Singer, eds. New York: St. Martin's Press, 1993.

Novak, M. A., and A. J. Petto, eds. *Through the Looking Glass: Issues of Psychological Well-Being in Captive Nonhuman Primates.* Washington, D.C.: American Psychological Association, 1991.

Novak, M. A., and S. J. Suomi. Psychological well-being of primates in captivity. *Am. Psychol.* 43: 765–773, 1988.

Paller, K. A., M. Kutas, and H. K. McIsaac. Monitoring conscious recollection via the electrical activity of the brain. *Psychol. Sci.* 6: 107–111, 1995.

Posner, M. I., and M. E. Raichle. *Images of Mind.* New York: American Scientific Books, 1994.

Rakover, S. S. *Metapsychology: Missing Links in Behavior, Mind and Science.* New York: Paragon House, 1990.

Rakover, S. S. Metapsychology: A pessimistic view of psychology? Psycoloquy, .95.6.35. metapsychology.9rakover, 1995.

Reinhardt, V. Improved handling of experimental monkeys. In *The Inevitable Bond: Examining Scientist-Animal Interactions*. Cambridge: Cambridge University Press, 1992.

Rollin, B. *The Unheeded Cry: Animal Consciousness, Animal Pain, and Science.* New York: Oxford University Press, 1989.

Rowan, A. N. The third R: Refinement. *ATLA—Alternatives Lab. Anim.* 23: 332–346, 1995.

Rowe, C., and T. Guilford. Hidden colour aversions in domestic chicks triggered by pyrazine odours of insect warning displays. *Nature* 383: 520–522, 1996.

Seligman, M. E. P., and J. L. Hager, eds. *Biological Boundaries of Learning.* New York: Appleton Century Crofts, 1972.

Sexton, O. J. Experimental studies of artificial Batesian mimics. *Behaviour* 15: 244–252, 1960.

Singer, P. *Animal Liberation*. New York: Avon, 1975.

Sjolander, S. Some cognitive breakthroughs in the evolution of cognition and consciousness, and their impact on the biology of language. *Evol. Cognit.* 1: 3–11, 1995.

Stricklin, W. R. Space as environmental enrichment. *Lab Anim.* 24(4): 24–29, 1995.

Terrick, T. D., R. L. Mumme, and G. M. Burghardt. Aposematic coloration enhances chemosensory recognition of noxious prey in the garter snake *Thamnophis radix. Anim. Behav.* 49: 857–866, 1995.

Tinbergen, N. *The Study of Instinct.* Oxford: Clarendon, 1951.

Tinbergen, N. On aims and methods of ethology. *Z. Tierpsych.* 20: 410–433, 1963.

Uexküll, J. von. Environment [*Umwelt*] and inner world of animals, C. J. Mellor and D. Gove, trans. In *The Foundations of Comparative Ethology,* G. M. Burghardt, ed. New York: Van Nostrand Reinhold, 1985 (reprinted from *Umwelt und Innenwelt der Tiere,* Berlin: Jena, 1909).

Uexküll, J. von, and G. Kriszat. *Streifzuge durch die Umwelten von Tieren und Menschen.* Berlin: Springer, 1934.

Underwood, G. The eye. In *Biology of the Reptilia,* vol. 2, C. Gans, ed. New York: Academic Press, 1970.

Warwick, C. Psychological and behavioural principles and problems. In *Health and Welfare of Captive Reptiles,* C. Warwick, F. L. Frye, and J. B. Murphy, eds. London: Chapman and Hall, 1995.

Whitaker, R. *Common Indian Snakes: A Field Guide.* New Delhi: Macmillan India Ltd., 1978.

Wilson, D. S., and E. Sober. Reintroducing group selection to the human behavioral sciences. *Behav. Brain Sci.* 17: 585–654, 1994.

7

MARC BEKOFF

Minding Animals

When a normal, naive dog receives escape/avoidance training in a shuttlebox, the following behavior typically occurs: At the onset of electric shock the dog runs frantically about, defecating, urinating, and howling until it scrambles over the barrier and so escapes from the shock. . . . However, in dramatic contrast to the naive dog, it [a dog who had received inescapable shock while strapped in a Pavlovian harness] soon stops running and remains silent until shock terminates . . . it seems to "give up" and passively "accept the shock."

Seligman, Maier, and Geer, 1968, p. 256

In one set of tests, the animals had been subjected to lethal doses of radiation and then forced by electric shock to run on a treadmill until they collapsed. Before dying, the unanesthetized monkeys suffered the predictable effects of excessive radiation, including vomiting and diarrhoea. After acknowledging all this, a DNA [Defense Nuclear Agency] spokesman commented: *"To the best of our knowledge, the animals experience no pain."*

Rachels, 1990, p. 132 (emphasis added)

From a humane point of view, there is no question that the lucky animals are those that are killed by people, whether it be by humane [*sic*] slaughter, a hunter, a car accident, or euthanasia by a humane organization or researcher.

Howard, 1994, p. 202

Actually, we lack evidence that game feel much initial pain when shot with a gun or arrow anymore than has been reported by many human military casualties or other traumatic injuries. *Also, animals do not suffer as much mental trauma as people.*

Howard, 1994, p. 204 (emphasis added)

> [A]n experiment that is a waste of time is also a waste of animals.
>
> Mann, Crouse, and Prentice, 1991, p. 13

> Recording and reporting such measures should be a routine part of any study using intrusive techniques, as the onus is on field workers to show that their methods have no impact, or at least an acceptable impact, on their study animals.
>
> Laurenson and Caro, 1994, p. 556

> I study foxes because I am still awed by their extraordinary beauty, because they outwit me, because they keep the wind and the rain on my face . . . because it's fun.
>
> Macdonald, 1987, p. 15

Introduction: A Privilege to Study Vulnerable Nonhuman Animals

Let me begin by saying that respecting all individuals' lives and treating them with dignity should direct our interactions with the animals with whom we share the planet. Within a traditional scientific milieu, what I offer may seem to be radical notions. However, if they stimulate informed debate, then the likelihood of change from all directions will be possible. After reading one of my papers on animal welfare, one of my colleagues asked me, "Hey, are you trying to put us out of business?" to which my reply was, "Of course not. I'm just trying to get you to think, even if you disagree with me." The quotations at the beginning of this essay adequately serve to make the rather unastonishing point that people do disagree about the sorts of permissible treatment to which nonhuman animals may be subjected by humans. I stress the word *nonhuman* rather than *subhuman* (e.g., Adler, 1993, p. 291; Björkqvist and Niemalä, 1992; Gallup, 1970, p. 87; Kennedy, 1992, p. 17), for the word *subhuman* suggests that animals are lower than humans, which is insulting to animals and also wrong in an evolutionary sense.

In a nutshell, my personal views, while certainly open to discussion and change, stem from numerous close encounters with a wide variety of nonhumans and humans. Their collective influence, which has resulted from their unselfish sharing of their lives and ideas with me, is clearly reflected in my views on animal well-being and the nature of their minds. These views include (1) taking the animals' points of view; (2) putting respect, compassion, and admiration for other animals first and foremost; (3) erring on the animals' side when I am uncertain about their feeling pain or suffering; (4) recognizing that almost all of the methods that are used to study animals, even in the field, are intrusions on their lives and that much research is fundamentally exploitive; (5) recognizing how misguided are "speciesistic" views concerning notions such as intelligence and cognitive or mental complexity for informing assessments of well-being (Bekoff, 1997, 1998a); (6) focusing on the importance of individuals; (7) appreciating individual variation and the diversity of the lives of different individuals in the worlds within which they live; (8) appealing to what some call questionable practices that have no place in the conduct of science, such as common sense and empathy; and (9) using broadly based rules of loyalty and nonintervention as guiding principles. Although I have always been concerned with animal well-being, I have not always applied the same standards of conduct to my own research. The bottom line is that we must dispense as rapidly as possible with ethical indifference to vulnerable animals who are unable to give direct consent for how they are used by humans. We must never forget that we are the animals' voices—we speak for them, like it or not! We must not only keep our minds open concerning new solutions to problems that have plagued us for a long time but also keep our hearts open to the plight of the animals for whom we speak. Of course, in the end, our lives also depend intimately on how we view and use other animals.

So, what I basically want to do here is to write about why I, an ethologist with strong interests in classical ethology (Allen and Bekoff, 1995a, 1995b; Bekoff, 1972, 1977a, 1991, 1995a), cognitive ethology and philosophy of mind (Allen and Bekoff, 1994, 1997; Bekoff, 1995b; Bekoff and Allen, 1992, 1997; Bekoff and Jamieson 1990a, 1990b; Jamieson and Bekoff, 1993), behavioral ecology (Bekoff, 1977b, 1995c; Bekoff and Wells, 1986), animal welfare and moral philosophy (Bekoff, 1993a, 1993b, 1994a, 1994b, 1994c, 1995d, 1995e, 1997, 1998a, 1998b; Bekoff and Hettinger, 1994; Bekoff and Jamieson, 1991, 1996; Bekoff et al., 1992; Jamieson and Bekoff, 1992), have arrived in the territory I now occupy concerning human-animal interactions. My primary objective is to show that ethology and different areas of philosophical inquiry reciprocally inform and motivate one another in the quest to get as many humans as possible to consider equally the interests of as many individual nonhumans as possible.

While I realize that I freely trespass into other disciplines and that some of my arguments are not fully developed in the way that philosophers might like them to be—in some instances, I am "talking" philosophy rather than "doing" philosophy—and also that there may be unanticipated places where my views may lead, I think that interdisciplinary exchange is valuable and necessary. I want to know what philosophers are thinking about cognitive ethology, and I think that it would be useful for philosophers to know what people like me are thinking about philosophical issues that are related to animal well-being; I can help them see how I view similar problems from a context with which most of them have had little or no direct experience. For example, I discovered the work of the philosopher Paul Taylor (1986) after I had written about certain aspects of loyalty and nonintervention. An understanding of the ways in which I arrived on similar (though not exactly the same) terrain could be helpful to him if he were to consider revising some of his ideas (for example, locating domesticated animals and others who are dependent upon humans more squarely in his biocentric ethic and using data from cognitive ethological studies for learning more about trust and deception). Simply put, we should strive to keep boundaries between disciplines semipermeable. Disciplinary arrogance will delay progress on important issues (Bekoff, 1994d), as might the fear of making mistakes (Orr, 1994).

I also want to stress how important are broad, comparative, and evolutionary studies of animal behavior. These types of endeavors will help us learn more about the individual animal's world and allow us to apply standards of conduct that are more in line with the needs of the individual members of the diverse taxa that are studied—how they live and use modalities other than vision. Humans are visual animals, and perhaps we need to pay more attention to the fact that animals can be helped and harmed by exposure to, for example, auditory or olfactory stimuli. I also reject the practice of speciesism, in which animals are treated according to the biological species to which they are assigned, and prefer the view that moral consideration of individuals (Bekoff and Gruen, 1993; Rachels, 1990) is of paramount importance in any debate about how humans view and treat their animals (as well as other humans). Rachels (1990, p. 173) favors the view he calls moral individualism, in which "the basic idea is that how an individual may be treated is to be determined, not by considering his group memberships, but by considering his own particular characteristics." Treating individuals differently "cannot be justified by pointing out that one or the other is a member of some preferred group, not even the 'group' of human beings" (p. 174). In my view, such a position can make the reintroduction of predators, for example, difficult to justify. Even if predators who had been eliminated by humans are given an opportunity to live where their ancestors had previously lived, prey animals who would not have been killed in the absence of reintroduction would now be killed. Human activities that result in the loss of lives must be given serious attention, for there seems to be little (but on my view there should be no) question that animals are a vulnerable class of individuals deserving protection (Shrader-Frechette, 1994).

Needless to say, it is important that we take our moral and ethical obligations to nonhumans as seriously as we deal with our moral and ethical obligations to other humans and to the world in which we live. Ethical issues are integral and legitimate parts of science. "Moral privatists" (Jamieson, 1985) who dispense with their moral and ethical obligations to nonhumans (White, 1990) are taking a position on matters, even though they seem ignorant that they are (Bekoff, 1991). These supposed amoralists are so detached from the world that they need sympathy and therapy or "help, or hope, not reasonings" (Williams, 1972, p. 1).

Finally, let me stress that this chapter has been terribly difficult to write, for at once it might seem that I am both self-serving—listing my and my close associates' papers—and trying to convince my colleagues that they should do as I say. It has also been a very frustrating experience; often I sit back and feel that I am wasting my time, for it should be obvious that we need to reconsider seriously what humans do to nonhuman animals. Allow me also to emphasize that I do not mean to be prescriptive, preachy, or simplistic, and let me apologize right now if that is how I come across. I suppose that if I offend some, then at least I have made them think. Furthermore, if some of what I write seems naive or "unscientific," this is how I really feel when I let my hair down.

I have asked myself repeatedly why I was asked to write a chapter such as this. Who could really care about what I have to say? Of course, I was honored and flattered by the request but scared and embarrassed as well. On the rare occasions when these sorts of essays are written by scientists, they are usually done by people near the ends of their careers; I hope that this does not apply to me! Furthermore, scientists rarely expose publicly their deep thoughts about human-nonhuman relationships and life in general. With respect to the question "Why me?" I have spent a considerable amount of time studying and thinking about animal behavior and animal welfare. While doing science, I have interacted with numerous philosophers with whom I share common interests in topics including animal cognition and ethics. I really enjoy collecting data, doing statistics, making tables, and drawing graphs, but I also like being connected to the animals I study.

General Background Assumptions: Having It Both Ways and Giving Animals the Benefit of the Doubt

First, I have no doubt that one can have it both ways. One can do "good science" and also respect the animals with whom one works. I will return to this topic later. Second, I assert that it is almost always wrong for humans intentionally to cause harm to any human or nonhuman animal or purposely to kill any human or nonhuman animal, including sea horses, ants, bees, worms, rats, mice, birds, cats, dogs, chickens, cows, and primates. While this statement may seem to be very strong, I believe that if it is used as a guiding principle, then it will force those who are in the position to use and to abuse animals to think deeply about what they are doing every time they make a decision to use an individual. Intentionally causing harm and possibly certain death in self-defense or to reduce further interminable suffering may be justified in certain circumstances. However, in general, when there is no reason to believe that human interference will lead to the cessation of further pain and suffering, it is immoral to cause harm or to take a life, and we have a prima facie duty not to do so. Nonhumans, like humans, may be harmed behaviorally, anatomically, or physiologically and not show overt responses until irreversible damage has been done. It is essential that all people who use animals know their subjects well so that they are able to assess if any harm has been done, even though it is not overtly obvious. There is no substitute for careful observation and description of what are considered to be "normal" or typical individuals.

I also assert that, when in doubt about the negative effects of intentional human action directed toward nonhumans, whether for purposes of research, education, amusement, or food, we should err on the side of animals. If we are not certain that an individual will not suffer when exposed to a given situation, we should not expose that individual to the given situation until we are certain that they will not suffer; we should presume the worst scenario and proceed from there (see also Bradshaw, 1990; Caplan, 1983). Some similar suggestions have also been offered for humans. For example, with respect to pain in infants, Fitzgerald (1987) claims that the question "Do infants feel pain?" is unanswerable and unhelpful when taken at face value. She notes (p. 346) that it is "better to assume that they do, take a step sideways and ask the question 'Can we measure pain responses in infants and are these measures sensitive to analgesics?'" The study of facial expression in infants seems to be one of the most promising measures of their pain responses. It may also be misleading to assume that anesthesia always causes loss of consciousness. This is not the case for humans (see Kulli and Koch, 1991), as evidenced by patient reports after surgery. I wonder how a nonhuman could possibly tell a human about pain after surgery?

It is also important to make serious attempts to take the animals' point of view and to try to discover answers to the fascinating question of how nonhumans interact in their own worlds and why they do so—what it would be like to be a particular individual from that individual's own perspective, not merely from our tainted view or from typological thinking about members of the same or closely related species. Limited time, expense, and methodological difficulties should not be used as excuses for producing hasty and misleading views on the ways in which nonhumans interact in their worlds, positions that might bear strongly on how they are treated. Others may not be like us because they are not one of us, and they may be different from other members of the same or closely related species because of genetic or acquired variation. While some animals seem to respond like humans to a wide variety of stimuli that are known to us to be pleasurable or painful, and arguments from analogy are often very convincing, we also know that many other animals process sensory information in nonhuman ways. Furthermore, it also is known that animals perform motor activities that are unlike those that humans typically perform. In these cases, arguments from analogy sometimes fail, but they do not always fail.

Some Personal Reflections

> At that point I was working with squid, and I think squid are the most beautiful animals in the world. And it just began to bother me. I began to have the feeling that nothing I could find out was worth killing another squid.
>
> (Ruth Hubbard, quoted in Holloway, 1995, p. 49).

My parents, although to some extent they still cannot figure out how I came to the profession that I am pursuing, tell me that I have always "minded" animals. Although I was not raised with animals, I used to ask about what they might be thinking or feeling as they went about their daily activities. In a nutshell, the phrase *minding animals* means caring for them, respecting them, feeling for them, and attributing minds (mental states and content) to individuals. Since I began working with animals, I have spent a lot of time pondering nonhuman–human relationships. Now, I often find myself obsessed with the terrible things that humans do to other animals with whom they share this planet. I recognize fully that most people who harm nonhumans for purposes of research, education, or amusement also bring some joy to some animals at other times. That nonhumans do not always suffer at the hands of humans seems to me to be a given that requires no elaboration. I find myself focusing on horror stories not because I am a pessimist whose glass is always half empty but rather because I think that it is more necessary to call attention to the incredible amount of pain and suffering that nonhumans experience at the hands of humans than to remind people of the good things that are done by most humans for nonhumans' benefits.

As time goes on, I find myself growing more and more "radical," and I frequently think it absurd that we should need laws to protect animals from humans and that a book such as this is even needed. I place the word *radical* in quotes because it also strikes me as perverse that to some *radical* means giving animals the benefit of the doubt with respect to their capacities to suffer and to experience pain, rather than referring to more permissive attitudes concerning the use of nonhumans by humans.

Basically, my early scientific training as an undergraduate and a beginning graduate student was grounded in what Rollin (1989) has called the "common sense of science," in which science is viewed as a fact-gathering, value-free activity. Of course, this outlook is not the case at all, but it took some time for me to come to this realization because of the heavy indoctrination concerning the need for scientific objectivity. With respect to the plight of the nonhumans who were used in classes or for research, there was little or no overt expression of concern for their well-being. Questions concerning morals and ethics rarely arose. When they did, they were invariably dismissed either by invoking what I have come to call "vulgar" or "facile" utilitarianism, in which suspected costs and benefits were offered from the human's point of view with no concern for the nonhuman's perspective, or by simply asserting that the animals really did not know, care, or mind (or whatever word could be used to communicate the animal's supposed indifference to) what was going on. Only once do I remember someone vaguely implying that something beneficial for the animal might come out of a research project.

After I took an undergraduate degree emphasizing anthropology and biology, I enrolled in a master's degree program in biology. I was interested in learning and memory and wrote a master's thesis on this topic. While pursuing this degree, two things happened that forever changed my perspective on how I should interact with animals. First, while taking a physiology course, it became clear to me that reductionist-mechanistic explanations of animal behavior were extremely narrow; flowcharts and input-output diagrams were too simple and impersonal. They also seemed to provide impersonal excuses for people to do the things they were doing to animals in the name of science. Second, one afternoon one of my professors calmly strutted into class announcing, while sporting a wide grin, that he was going to kill a rabbit for an experiment by using a method named after the rabbit itself, namely, a "rabbit punch." He proceeded to kill the rabbit by chopping it with the side of his hand and breaking its neck. I was astonished and sickened by the entire spectacle. I refused to partake in the laboratory exercise and also decided that what I was doing at the time was simply wrong for me. I began to think seriously about alternatives. I enjoyed science, but I suspected there were other ways to do science that incorporated respect for animals and allowed for individual differences among scientists concerning how science is done.

Because of my developing interests in animal behavior—a field that was considered by most at the time to be a soft science (or akin to stamp collecting, de Solla Price, 1960)—and my interests in neurobiology, I entered a graduate program in which I could tie together these two fields. I realized that I would have to kill (some would say euthanize or sacrifice) animals as part of any research in this discipline, but for some reason that I cannot clearly recollect I decided to put that out of my mind and enroll in this program. My research centered on vision in cats. I truly enjoyed and was challenged by trying to figure out how cats saw their world. However, once the killing of experimental animals began, I also truly hated having to kill the animals to localize lesions that were made in various parts of their brains. For one reason or another, I became a good executioner and wound up killing others' cats as well.

One morning, I woke up very disturbed about the whole thing and decided that I could not continue to kill cats. I was especially tormented by thinking about the eyes of the cats as they were being prepared to be killed. What an undignified end to a life! I simply did not want to kill any animals as part of my research. I just could not justify this murder by using any form of utilitarianism. I (and three medical students) also refused to partake in some physiology experiments that used dogs, and, to our amazement, we were excused without prejudice from doing so, although the distinguished professor could not understand why we did not want to kill the dogs; he asserted that they would have died in an animal shelter anyway. To his credit, though, he remained true to his word, and I applaud his permissiveness and his open-mindedness. At the end of the term, despite these reprieves, I left this program because I could not do the research that I wanted to without killing animals or being responsible for their deaths.

While I was looking for another graduate program, I came across work on the development of behavior in various canids (coyotes, wolves, dogs, foxes) that Michael W. Fox was doing at Washington University in St. Louis. After long talks with him about animal behavior and also my concerns for animal welfare and the cold, unemotional views of so many scientists, we began working together on a variety of projects, one of which became the subject of my doctoral dissertation. Mike and I continued to have numerous serious discussions about animal behavior and animal welfare, pondering questions such as "What it is like to be an animal in its world?"; "How do we know when and if an animal is suffering?"; and "What can we do to minimize or abolish the abuse of nonhumans by humans?" A question that haunted me then as now is "How can people do the terrible things they do to nonhumans?" Being a child of the 1960s, one easy answer was that, given that it was so easy for humans to kill other humans, especially those who lived in foreign lands, how could we reasonably think that animals would be spared this wanton abuse? I also pondered, "What kind of people could detach themselves from an animal that is obviously suffering, and why might a person choose to do this?" I believed then, and still do, that people have a choice as to what types of research they are going to pursue and that, for some, harming and killing animals simply is not disturbing enough to the point that they will stop doing what they are doing. Even in my own research, I came to the realization that I was doing things that I could no longer tolerate. For example, after I did some studies in which I intentionally allowed coyotes to kill chickens and mice in staged encounters (Vincent and Bekoff, 1978), I stopped when I realized there was something grossly inhumane about the work. I would not do this type of research again. If I have to forgo learning something because to do so entails killing animals, then so be it.

After completing my doctoral dissertation and doing some postdoctoral work, I was offered a faculty position at the University of Colorado in Boulder and had an opportunity to do some research in Antarctica on the behavior of Adélie penguins (Bekoff, Ainley, and Bekoff, 1979). In 1975, I began a long field project on the social ecology and behavior of coyotes (summarized in Bekoff and Wells, 1986). I also wrote some papers in the mid-1970s concerning animal behavior and animal welfare (Bekoff, 1976) and a book review (Bekoff, 1980) that, on reflection, showed me that I really had been thinking about these issues for a long time.

My personal notes from my days spent in Antarctica and in the field studying coyotes and birds consistently question what types of research, including field studies of behavior and behavioral ecology, could be justified. I recognized that field work, like behavioral research on captive animals, could be a significant intervention into their lives, although some field workers are very chauvinistic about how their research is less disruptive than is work on captive animals. I thought about how trapping and marking individuals— and simply being there—influenced the lives of the animals we were studying (Bekoff, 1995d). My thoughts on these matters were incorporated into my long-term study of coyotes. While doing this research, we took all necessary precautions to minimize harm when trapping and tagging coyotes, but we still had one injury that we could assign to our trapping efforts. I would find it difficult to justify doing this work again, although I know that to answer the questions that interested us trapping and marking were necessary.

I continued to ponder these questions while I studied (and continue to study) western evening grosbeaks (Bekoff et al., 1989; Bekoff and Scott, 1989; Bekoff, 1995c). My experience with these birds has made me even more sensitive to why humans study animals in the first place and what types of research are permissible. For example, I asked myself whether it is really important (1) to be able to follow identified individuals through as much of their lives as possible to know whether brothers were helping sisters or whether sisters were helping sisters or mothers to rear young, (2) to know whether individuals were predisposed to provide aid to kin who were engaged in potentially injurious fights or to warn them preferentially that a potential predator was present, or (3) to learn about infanticide (Bekoff, 1993a). I concluded that acquiring reliable information for these and other questions is important, as long as the interventions that were necessary were minimally disruptive or harmful. But I fully realized that I was making this decision for the animals, using my own criteria for deciding what was "minimally disruptive" or "harmful." This thought was, and remains, disturbing to me (see also Farnsworth and Rosovsky, 1993).

My laboratory and field experience showed me that all behavioral research intervenes, even that which appears to be simple observation. This fact of the matter must be taken seriously by all researchers. As I continued my empirical studies, I found myself more and more interested in how my interests in evolutionary biology and cognitive ethology could be tied together with my interests in moral philosophy, philosophy of mind, and animal welfare. Because of prominent scientists' recent accusations about the presumed intentions and characteristics of those who are interested in animal rights and animal welfare, it is essential to stress that I am not (and have never been) antiscience (Bernstein, 1989), an anti-intellectual, or a Luddite (Bernstein, 1989; Nicoll and Russell, 1990), and I certainly do not want to halt all animal research (Mason, 1990). The combative tone of Nicoll and Russell's title

("Ammunition for Counter Offensive by Scientists")—you would think they were fighting a world war—and their conclusion that scientists are being selectively targeted by proponents of animal rights are based on an analysis of some of the available written literature on animal rights and animal welfare. They do not consider the other constructive ways in which those interested in animal welfare, in other areas as well as the sciences, communicate their ideas.

My close relationship with a number of philosophers showed me that it really was possible for someone to do science, to question scientific practices, and to consider animal welfare seriously. A person does not have to be antiscience or anti-intellectual to question how science is done; it just seems that an unwritten rule of science mandates that scientists should not question science. One of my first goals was to come to terms with many different and extremely complex ideas in moral and ethical theory. I needed to understand differences and similarities (if any) among schools of thought favoring contractarianism, various types of utilitarianism, rights, and other moral systems. I also had to learn what terms such as *rights, duties, obligations,* and *interests* meant to the people who used them.

Another goal was to get individuals' eponymous positions correct, for the different major players in discussions of animal welfare are not saying the same things. For example, Peter Singer's (1990) utilitarian views do not make him an animal rightist, as is so often suggested, and Singer's stance is quite different from that of Tom Regan's (1983) rights approach (see, for example, the exchange between Singer [1985] and Regan [1985] in the *New York Review of Books*). Nonetheless, their views are often presented as being one and the same. The more I learned about their (and others') differences, the more incredulous I became when I realized that intellectually brilliant people, who seemed to be rational, disagreed on what could have been obvious points of agreement about whether at least certain animals are conscious, whether animals suffered and experienced pain, and what human obligations are due to nonhumans.

One thought that I continue to wrestle with is moderation with respect to the use of animals by humans (Bekoff and Jamieson 1991, pp. 24ff.; Finsen, 1990). Is it a cop-out? Is a moderate view a hybrid view that is doomed to failure? A moderate view can easily become a self-serving compromise, for when I conclude that something is permissible, it is because I have entered the figures for costs and benefits into a utilitarian calculus, for example, and I can make it come out just about any way I please. Correctly assigning costs and benefits is usually tenuous at best. In most cases, we can only make educated guesses, and these might not really tell the true story with respect to animal suffering and pain and the value of whatever kinds of data are collected and knowledge is obtained.

So, should I take a stronger position that centers on stringent prescriptive codes of conduct? Difficult issues rarely are so cut and dried, but perhaps if we are to make serious attempts to stop animal abuse, hard-and-fast rules need to be used. Right now, I simply do not have an answer. I consider myself to be a moderate with respect to the use of animals in research, although I favor stringent restrictions. My moderate stance permits some animal research to be done after it is rigorously scrutinized. However, I also believe that some types of behavioral research should not be permitted at all, including, but not limited to, the use of inescapable shock, various forms of social deprivation (see Stephens, 1986), various types of physical restraint, extreme sensory deprivation, extreme starvation, artificially staged encounters to study aggressive or predator-prey relationships (especially those in which animals are unable to escape from one another), some experimental studies of infanticide in which animals are injured or killed, and studies in which animals are castrated or are rendered unable to vocalize, to see, to hear, or to feel tactile stimuli.

I fully recognize that I am not providing strict guidelines, for I favor looking at studies on a case-by-case basis. For example, in some instances, such as when they are being prepared for reintroduction to wild habitats, animals may have to be subjected to stressful situations that replicate conditions in their natural environments (Beck and Castro, 1994). Although these sorts of treatment may seem abusive, they may also be necessary to ensure the survival and well-being of individuals who are to be released. Furthermore, although some studies of infanticide may indeed be extremely abusive and compromise both the well-being of individuals who are intentionally killed by experimenters and those who suffer and die because of the removal of protective adults (e.g., Emlen, Demong, and Emlen, 1989; for discussion, see Bekoff, 1993a), there are ways in which studies of infanticide can be done more humanely (e.g., Perrigo et al., 1989). Certainly, it seems as if Emlen and colleagues' work on infanticide in the polyandrous wattled jacana could have been conducted with the well-being of the animals in mind without compromising the results.

Now, having made these claims about some lines of research that I think are extremely difficult to justify, I want to stress here, as I have before (Bekoff and Jamieson, 1991, p. 26), that reasonable people can disagree about some particular lines of research. However, reasonable people cannot disagree about the necessity for reforming our practices with respect to animal use and the changes of outlook that may be required to do so. A moderate stance can take a hard line, and no one wants to be told that their or their colleagues' work is immoral. Of course, people with good intentions and character can act wrongly. However, name-calling and finger-pointing will not be useful in changing either how people view animals or how they use them.

Thinking about my position still keeps me awake at night, for there are problems with it. Maybe the passage of time will result in clarity on at least some of these difficult issues. Although space does not allow me to go into details, I have found Francione's (1994) discussion of the contrasts between legal welfarism—which, among other things, characterizes animals as human property and permits conduct that maximizes the value of animal property—and legal rights views, which do not, to be very stimulating. Francione worries that "as long as animal 'interests' are being assessed within a system that accords legal rights to humans but not to animals, then animals will virtually always lose any purported 'balancing' of human and animal interests" (p. 770).

Another problem is consistency, which is especially troublesome when we consider where to draw the line between what type of animal use is permissible and what is not. When I think about ethical and moral issues and how humans and nonhumans ought to interact, I find it difficult to come up with a consistent train of thought. Sometimes I find myself concluding that all animal research should be terminated, along with the use of animals in education, for amusement, and for food.

At other moments, when, for example, I think about research that can benefit animals as well as humans, I retreat to my principled and highly restrictive moderate stance that centers on (1) accepting that it is almost always wrong to harm other animals, (2) assuming that all individuals, regardless of species, suffer to some degree unless there is incontrovertible evidence to the contrary, (3) using observations of self-regarding behavior in which an individual avoids either what we would call a noxious stimulus or a situation that we call noxious based on what we know about the animal's sensory or social environments, (4) using as few animals as possible only when there are no alternatives (despite Ng's [1995] claim that a reduction in animal suffering may be achieved by doing more, rather than less, animal research), (5) using the most humane methods known, (6) being certain that the well-being of all animals is given serious attention after they have been used, (7) telling all potential readers of scientific papers how animals were negatively affected by the research so that others could avoid making the same mistakes (Morton, 1992), and (8) recognizing that humans are necessarily anthropocentric and that the animal's point of view can never be totally assimilated into the utilitarian calculus, regardless of how right-minded individual people are. These strictures mandate that animals should be used only as a last resort. In the best of all possible worlds, animals would not be used at all. Limited time, money, and energy (and motivation) are not excuses for using animals when alternatives are available or can be developed.

Some Difficult and Disturbing Questions: Peter, Paul, Mary, and Tom

> If we, in the western world, see a peasant beating an emaciated old donkey, forcing it to pull an oversize load, almost beyond its strength, we are shocked and outraged. But, taking an infant chimpanzee from his mother's arms, locking him into the bleak world of the laboratory, injecting him with human diseases—this, if done in the name of Science, is not regarded as cruelty. Yet in the final analysis, both donkey and chimpanzee are being exploited and misused for the benefit of humans.
>
> Why is one any more cruel than the other? Only because science has come to be venerated, and because scientists are assumed to be acting for the good of mankind, while the peasant is selfishly punishing a poor animal for his own gain. In fact, much animal research is self-serving too—many experiments are designed in order to keep the grant money coming in. (Goodall, 1990, pp. 249–250)

> They were two human primates carrying another primate. One was the master of the earth, or at least believed himself to be, and the other was a nimble dweller in trees, a cousin of the master of the earth. (Preston, 1994, pp. 56–57)

Let me emphasize that studying nonhuman animals is a privilege. We must take this privilege seriously. While some of my colleagues may conclude that I am trying as hard as I can to end my career as a scientist and also to hamper their own work, I am not. I simply believe that these (and other) questions demand serious consideration. Many principles have been proposed that perhaps could guide us in our treatment of animals: utilitarian ones, rights-based ones, interests-based ones, and so forth. Using

Peter (Singer, 1990), Paul (Taylor, 1986), Mary (Midgley, 1983), and Tom (Regan, 1983) as our guides seems a good idea at the moment, although, of course, all have their critics (e.g., Nicoll and Russell, 1990; Russell and Nicoll, 1996, and references therein; see also Singer, 1996). Scientists often operate on the basis of implicit principles and guidelines that are often not discussed. All of these principles need to be brought out into the open and explicitly debated. Foremost in any deliberations about other animals must be deep concern and respect for their lives and for the worlds in which they live—respect for who they are in their worlds and not respect motivated by who we want them be in our anthropocentric scheme of things (see also Westra's [1994] important discussion of integrity). As Taylor (1986) notes, a switch away from anthropocentrism to biocentrism, in which human superiority comes under critical scrutiny, "may require a profound moral reorientation" (p. 313). So be it.

Here are some questions I ponder. I include them because I hope that deep consideration of these sorts of questions will result in better treatment of animals in the future. How can people subject animals to inescapable shock or sew their eyelids shut or poison or shoot them, for example, in the name of science? How can people be so objective as to write about screaming dogs, whining mice, self-mutilating monkeys, or starved snakes without themselves suffering greatly? What kind of person is able to do harmful things to animals and then sleep at night? What kind of people can harm a dog, for example, "at work" and then go home at night and play with their companion dog and tell everyone how smart or witty or affectionate their companion is? Why are some people offended or threatened by common sense and anthropomorphic approaches to the interpretation and explanation of behavioral data, especially when there are moral consequences? As Sprigge (1988) notes: "The prime reason why it is wrong to torture a dog is just the same as the prime reason why it is wrong to torture a human being, namely that it hurts, to put it mildly" (p. 219). Even if common sense is wrong some of the time, as are scientifically based explanations, they are also right some of the time.

Because studies of animal minds and cognitive ethology inform matters of welfare, I also ask why some people apply a different standard for research in cognitive ethology than they do for other sciences. Why do some people dismiss the likelihood that many animals have rich intentional lives or suffer because it is hard to perform studies that produce solid results? How can one deny that a dog has some beliefs about what it is doing, even if its beliefs are not like ours? Biologists do not agree on concepts such as adaptation, fitness, or even evolutionary mechanisms. But they do not dismiss these ideas because they are difficult to study. We readily accept evolutionary continuity in physiology and anatomy; why not in behavior? Can we really believe that we are the only species with feelings, beliefs, desires, goals, expectations, the ability to think, or the ability to think about things?

Toward a Deep Ethology: One Can Have It Both Ways

> One of the great dreams of man must be to find some place between the extremes of nature and civilization where it is possible to live without regret. . . . (Barry Lopez, quoted in McKibben, 1995, p. 80)

> Certainly it seems like a dirty double-cross to enter into a relationship of trust and affection with any creature that can enter into such a relationship, and then to be a party to its premeditated and premature destruction. (Johnson, 1991, p. 122)

> If we conclude that chimpanzees are conscious, we must then confront the ethics of our treatment of such animals in captivity and in the remaining wild. (Jolly, 1991, p. 231)

Clearly, the issues with which I have dealt here—and I must stress that many others have been ignored because of space limitations—are extremely difficult ones that do not submit to easy solutions. The more I study nonhuman animals, the more difficult it is to imagine doing anything harmful to them (see also Ulrich, 1989; Gluck, 1997). Naming and bonding with the animals I study is one way to respect them (Bekoff, 1994c; see also Davis and Balfour, 1992). The ethicist Arthur L. Caplan (personal communication, April 1, 1991) has shared with me his feelings about this issue. He notes:

> It does seem to me to be one of the great ironies of the debate over the moral status of experimentation with animals that the more observation and experimentation that is done, the harder it is to avoid drawing the conclusion that some animals have mental lives and therefore moral standing that must restrain the kinds of observation and experimentation that is done to them.

Surprises are always forthcoming concerning the cognitive skills of nonhumans, and people who write about animal issues must be cognizant of these findings. I do not see how any coherent thoughts about the moral and ethical aspects of animal use could be put forth without using biological-evolutionary, ethological, and philosophical information. Ethologists need to read philosophy, and philosophers must not only read ethology but also watch animals (as Daniel Dennett [1988] did when he studied with Dorothy Cheney and Robert Seyfarth [see their 1990 book]).

I believe that a "deep ethology" is needed to make people more aware of what they do to nonhumans and to make them aware of their moral and ethical obligations to animals (Bekoff and Jamieson, 1991). I use the term *deep ethology* to convey some of the same general ideas that underlie the deep ecology movement (Tobias, 1988), in which people recognize that they not only are an integral part of nature but also have unique responsibilities to nature. Most people who think deeply about the troubling issues surrounding animal welfare would agree that the use of animals in research, education, and amusement and for food needs to be severely restricted. Our unique responsibilities to the more-than-human world (Abram 1996) mandate, in my view, that a noninterventionist policy should be ours in the future. We need to accept that many nonhuman animals experience pain and suffering, even if it is not the same sort of pain and suffering that humans or even other nonhumans experience. We must also rigorously determine just how invasive are most animal studies (Barnett and Hemsworth, 1990; Field, Shapiro, and Carr, 1990; Report of the Laboratory Animal Science Association Working Party, 1990; Shapiro and Field, 1987).

In our quest to change human attitudes toward nonhuman animals, our perceptions of animals must change so that we view them as subjects rather than as objects. Many intermingled layers need to be considered, and unidimensional thinking characteristic of "flat-earth" (human chauvinists and speciesists) inhabitants will have to be replaced by multidimensional thinking. Perhaps perceptions of, and feelings about, how animals are treated will change if people have opportunities to observe animals performing activities that clearly indicate that they are capable of feeling pain and suffering and also capable of thinking and of having beliefs, desires, and expectations. Furthermore, via films or actual visits, people could see what goes on in different research, educational, and amusement facilities, especially those that both strong supporters and weak supporters of animal welfare agree are "good" (animal welfare is seriously taken into consideration and something is done about it) or "bad" (animal welfare is not taken seriously, which is obvious to most, if not all, outside observers).

In efforts to foster strong and long-lasting changes, we probably should start with young children (for a good example, see Kleiman et al. [1986], who also write about education programs in general). The common sense with which they approach the world is refreshing, as are their well-developed philosophical skills (Matthews, 1984, 1994; see also Dickinson, 1988). Children also naturally seem to sympathize and to empathize with animals (Bly, 1992). As McGinn (1991, p. 16) notes, "Children are the most natural friends of animals, and paying them more respect might be the best way to get animals liberated." Of course, children can also be extremely cruel to animals and to other children. The point here, as I see it, is that it might be easier to make life-long changes in individuals' attitudes toward nonhumans (and other humans) if codes of conduct are presented as early in life as possible, before irreversible attitudes have been established. Books in the popular press also are helpful in getting across messages about animal welfare to a wide audience (e.g., Hall, 1995; Hiaasen, 1991; McQuillan, 1993; Preston, 1994; Quinn, 1992; Tobias, 1994).

Although arguments about rights and duties and obligations certainly are essential and highlight the important and complex issues that surround discussion about animal welfare and protection, shedding philosophical robes and adopting a common sense approach to many of the issues, especially how we view the cognitive skills of nonhumans and their pain and suffering, will make a better world in which humans and nonhumans can live compatibly. Of course, our common sense intuitions about pain, suffering, and animal cognition must be combined with reliable empirical data, of which there are plenty. Thus, I am concerned by claims that cognitive explanations have yet to prove their worth when compared with behavioristic explanations of the behavior of nonhumans. Colgan (1989, p. 67) eschews cognitive explanations and claims that they are no advance over "the anecdotalism and anthropomorphism which characterized ethology a century ago." Sebeok (1991, p. xii) also claims that the problem area of cognitive ethology "has not progressed much beyond anecdotal evidence since the publication of the work of G. J. Romanes, in 1892." Unfortunately, both Colgan and Sebeok ignore a plethora of data that had been collected before their claims were made (for references, see Allen and Bekoff, 1997; Bekoff and Allen, 1997; Byrne, 1995; Ristau, 1991). What people believe about the cognitive capacities of nonhumans informs their thinking about animal welfare; different views dispose a person to look at animals in particular ways (Beauchamp, 1992; Bekoff, 1994a). Ascribing intentionality and other cognitive abilities to animals is not moot if there are moral consequences, and there are. Nonetheless, I maintain that pain and suffering and not cognitive abilities should be the major issues guiding the use of nonhuman animals by humans.

People of all perspectives on animal welfare have to listen to one another. We will get nowhere without dialogue. "Them" versus "us" arguments and attacks (e.g., Russell and Nicoll [1996]; for reply, see Singer [1996]) will maintain the unfortunate division that has arisen in the exchanges among proponents of different views.[1] Humans who are deeply concerned with animal welfare should not be dispensed with as antiscience or anti-intellectual crazies, and people who are more permissive about animal welfare should not be dispensed with as necessarily bad people who are seriously misguided.

Today's students will live and work in a world in which science increasingly will not be seen as a self-justifying activity but as another human institution whose claims on the public treasury must be defended. Now more than ever, students must understand that questioning science is not antiscience or anti-intellectual and that asking how humans should interact with animals is not a demand that humans never use animals. Questioning science will make for better, more responsible science, and questioning the ways in which humans use animals will make for more informed decisions about animal use. By making such decisions in an informed and responsible way, we can help ensure that we will not repeat the mistakes of the past and that we will move toward a world in which humans and other animals can peaceably share the resources of a finite planet. Perhaps the enterprise of science needs to be redefined to include subjectivity and feelings. This move will make science an ally rather than a foe in our dealings with nonhuman animals.

The problems we face are complex and also novel. Keeping open minds—and, more important, open hearts—is essential. Facile attempts to dance or shuffle around difficult and unsavory situations "in the name of science" or within the constraints of "the scientific method" (as if there is one method) are not going to work in the future. We need to bite the moral bullet, but the moral truth is often a bitter pill to swallow.

Conclusions

Despite the long road to travel, change is in the wind. We and the animals we use should be viewed as partners in a joint venture. No one can be an island in this intimately connected universe. Furthermore, we should not continue being at war with the world. As Nobel laureate Barbara McClintock notes, we must have a feeling for the organisms with whom we are privileged to work. Thus, bonding with animals and calling animals by name are steps in the right direction. It seems unnatural for humans to resist developing bonds with the animals who they study. By bonding with animals, we should not fear that the animals' points of view will be dismissed. In fact, bonding will result in a deeper examination and understanding of the animals' points of view, and this knowledge will inform further studies on the nature of human-animal interactions. If we forget that humans and other animals are all part of the same world, and that humans and animals are deeply connected at many levels of interaction, when things go amiss in our interactions with animals, as they surely will, and animals are set apart from and inevitably below humans, I feel certain that we will miss the animals more than the animal survivors will miss us. The interconnectivity and spirit of the world will be lost forever, and these losses will leave a severely impoverished universe.

ACKNOWLEDGMENTS This essay is dedicated to my parents, who, perhaps without knowing it, raised me to view animals and humans as deserving equal respect, although I certainly have not always behaved in this manner, despite my efforts to do so. My parents also taught me that to err is to be human. I thank Marjorie Bekoff Morse, Ann Wolfe, Susan Townsend, Colin Allen, Kenneth Shapiro, Lynette Hart, and two anonymous reviewers for comments on an earlier draft of this chapter and for discussing many of the issues in it.

Note

1. It should be noted that Russell and Nicoll revised the prepublication draft of their 1996 paper after they discovered they were in error about one of their major beliefs concerning Singer's intention to mislead readers of his book, *Animal Liberation.* For discussion, see Bekoff (1994e). Cardarelli (1994) also points out logical errors in Nicoll and Russell's thinking.

References

Abram, O. *The Spell of the Sensuous: Perception and Language in a More-Than-Human World.* New York: Pantheon Books, 1996.

Adler, M. J. *The Difference of Man and the Difference It Makes.* New York: Fordham University Press, 1993.

Allen, C., and M. Bekoff. Intentionality, social play, and definition. *Biol. Philos.* 9: 63–74, 1994.

Allen, C., and M. Bekoff. Biological function, adaptation and natural design. *Philosophy of Science* 62: 609–622, 1995a.

Allen, C. and M. Bekoff. Natural design, function, and animal behavior: Philosophical and ethological considerations. *Perspect. Ethol.* 11: 1–46, 1995b.

Allen, C., and M. Bekoff. *Species of Mind: The Philosophy and Biology of Cognitive Ethology.* Cambridge, Mass.: MIT Press, 1997.

Barnett, J. L., and P. H. Hemsworth. The validity of physiological and behavioural measures of animal welfare. *Appl. Anim. Behav. Sci.* 25: 177–187, 1990.

Beauchamp, T. L. The moral standing of animals in medical research. *J. Law Med. Health Care* 20: 7–16, 1992.

Beck, B. B., and M. I. Castro. Environments for endangered primates. In *Naturalistic Environments in Captivity for Animal Behavior Research,* E. F. Gibbons, E. J. Wyers, E. Waters, and E. W. Menzel Jr., eds. Albany, N.Y.: SUNY Press, 1994.

Bekoff, M. The development of social interaction, play, and metacommunication in mammals: An ethological perspective. *Q. Rev. Biol.* 47: 412–434, 1972.

Bekoff, M. The ethics of experimentation with non-human subjects: Should man judge by vision alone? *Biologist* 58: 30–31, 1976.

Bekoff, M. Social communication in canids: Evidence for the evolution of a stereotyped mammalian display. *Science* 197: 1097–1099, 1977a.

Bekoff, M. Mammalian dispersal and the ontogeny of individual behavioral phenotypes. *Am. Naturalist* 111: 715–732, 1977b.

Bekoff, M. Review of M. W. Fox, Returning to Eden: Animal rights and human responsibility. *BioScience* 31: 533, 1980.

Bekoff, M. Animal ethics reconsidered. *Hastings Center Report* 21 (September–October): 45, 1991.

Bekoff, M. Experimentally induced infanticide: The removal of birds and its ramifications. *Auk* 110: 404–406, 1993a.

Bekoff, M. Common sense, cognitive ethology and evolution. In *The Great Ape Project: Equality Beyond Humanity,* P. Cavalieri and P. Singer, eds. London: Fourth Estate, 1993b.

Bekoff, M. Cognitive ethology and the treatment of nonhuman animals: How matters of mind inform matters of welfare. *Anim. Welfare* 3: 75–96, 1994a.

Bekoff, M. Animal welfare. *Am. Biol. Teach.* 56: 391–392, 1994b.

Bekoff, M. Should scientists bond with the animals whom they use? Why not? *Int. J. Comp. Psychol.* 7: 1–9, 1994c.

Bekoff, M. But is it research? What price interdisciplinary interests? *Biol. Philos.* 9: 249–252, 1994d.

Bekoff, M. Animal welfare. *Nature* 371: 99, 1994e.

Bekoff, M. Play signals as punctuation: The structure of social play in canids. *Behaviour* 132: 419–429, 1995a.

Bekoff, M. Cognitive ethology and the explanation of nonhuman animal behavior. In *Comparative Approaches to Cognitive Science,* J.-A. Meyer and H. L. Roitblat, eds. Cambridge, Mass.: MIT Press, 1995b.

Bekoff, M. Vigilance, flock size, and flock geometry: Information gathering by western evening grosbeaks. *Ethology* 99: 150–161, 1995c.

Bekoff, M. Marking, trapping, and manipulating animals: Some methodological and ethical considerations. In *Wildlife Mammals as Research Models: In the Laboratory and Field,* K. A. L. Bayne and M. D. Kreger, eds. Greenbelt, Md.: Scientists Center for Animal Welfare, 1995d.

Bekoff, M. Naturalizing and individualizing animal well-being and animal minds: An ethologist's naivete exposed? In *Wildlife Conservations, Zoos, and Animal Protection: A Strategic Analysis,* A. Rowan, ed. Grafton, Mass.: Tufts Center for Animals and Public Policy, 1995e.

Bekoff, M. "Do dogs ape?" or "do apes dog?" and does it matter: Broadening and deepening cognitive ethology. *Anim. Law.* 3: 13–23, 1997.

Bekoff, M. Deep ethology, cognitive ethology, and the Great Ape/Animal Project: Expanding the community of equals. In *Applied Ethics in Animal Research.* J. Gluck and B. Orlans, eds. West Lafayette, Ind.: Purdue University Press, 1998a.

Bekoff, M. (ed.). *Encyclopedia of Animal Rights and Animal Welfare.* Westport, Conn.: Greenwood Publishing Group, Inc., 1998b.

Bekoff, M., D. Ainley, and A. Bekoff. The ontogeny and organization of comfort behavior in Adélie penguins, *Pygoscelis adélie. Wilson Bull.* 91: 255–270, 1979.

Bekoff, M., and C. Allen. Intentional icons: Towards an evolutionary cognitive ethology. *Ethology* 91: 1–16, 1992.

Bekoff, M., and C. Allen. Cognitive ethology: Slayers, skeptics, and proponents. In *Anthro-*

pomorphism, Anecdote, and Animals: The Emperor's New Clothes? R. W. Mitchell, N. Thompson, and L. Miles, eds. Albany, N.Y.: SUNY Press, 1997.

Bekoff, M., and L. Gruen. Animal welfare and individual characteristics: A conversation against speciesism. *Ethics Behav.* 3: 163–175, 1993.

Bekoff, M., L. Gruen, S. E. Townsend, and B. E. Rollin. Animals in science: Some areas revisited. *Anim. Behavior* 44: 473–484, 1992.

Bekoff, M., and N. Hettinger. Animals, nature, and ethics. *J. Mammal.* 75: 219–223, 1994.

Bekoff, M., and D. Jamieson, eds. *Interpretation and Explanation in the Study of Animal Behavior.* Vol. 1: *Interpretation, Intentionality, and Communication.* Boulder, Colo.: Westview Press, 1990a.

Bekoff, M., and D. Jamieson, eds. *Interpretation and Explanation in the Study of Animal Behavior.* Vol. 2: *Explanation, Evolution, and Adaptation.* Boulder, Colo.: Westview Press, 1990b.

Bekoff, M., and D. Jamieson. Reflective ethology, applied philosophy, and the moral status of animals. *Perspect. Ethol.* 9: 1–47, 1991.

Bekoff, M., and D. Jamieson. Ethics and the study of carnivores. In *Carnivore Behavior, Ecology, and Evolution,* J. L. Gittleman, ed. Ithaca, N.Y.: Cornell University Press, 1996.

Bekoff, M., and A. C. Scott. Aggression, dominance, and social organization in evening grosbeaks. *Ethology* 83: 177–194, 1989.

Bekoff, M., A. C. Scott, and D. C. Conner. Ecological analyses of nesting success in evening grosbeaks. *Oecologia* 81: 67–74, 1989.

Bekoff, M., and M. C. Wells. Social ecology and behavior of coyotes. *Adv. Study Behav.* 16: 251–338, 1986.

Bernstein, I. S. Breeding colonies and psychological well-being. *Am. J. Primatol.* 1(Suppl.): 31–36, 1989.

Björkqvist, K., and P. Niemelä, eds. *Of Mice and Women: Aspects of Female Aggression.* New York: Academic Press, 1992.

Bly, R. *Iron John.* New York: Vintage Books, 1992.

Bradshaw, R. H. The science of animal welfare and the subjective experience of animals. *Appl. Anim. Behav. Sci.* 26: 191–193, 1990.

Byrne, R. *The Thinking Ape: Evolutionary Origins of Intelligence.* New York: Oxford University Press, 1995.

Caplan, A. L. Beastly conduct: Ethical issues in experimentation. *Ann. N. Y. Acad. Sci.* 406: 159–169, 1983.

Cardarelli, N. Letter to the editor. *Ohio J. Sci.* 94: 38–39, 1994.

Cheney, D., and R. Seyfarth. *How Monkeys See the World: Inside the Mind of Another Species.* Chicago: University of Chicago Press, 1990.

Colgan, P. *Animal Motivation.* New York: Chapman and Hall, 1989.

Davis, H., and D. Balfour, eds. *The Inevitable Bond: Examining Scientist-Animal Interactions.* New York: Cambridge University Press, 1992.

de Solla Price, D. J. Review of I. Asimov, *The Intelligent Man's Guide to the Universe. Science* 132: 1830–1831, 1960.

Dennett, D. C. Out of the armchair and into the field. *Poetics Today* 9: 205–221, 1988.

Dickinson, P. *Eva.* New York: Dell, 1988.

Emlen, S. T., N. J. Demong, and D. J. Emlen. Experimental induction of infanticide in female wattled jacanas. *Auk* 106: 1–7, 1989.

Farnsworth, E. J., and J. Rosovsky. The ethics of ecological field experimentation. *Conserv. Biol.* 7: 463–472, 1993.

Field, P. B., K. Shapiro, and J. Carr. Invasiveness of experiments conducted by leaders of psychology's animal research committee (CARE). *Psychol. Ethical Treatment Anim.* 10: 1–10, 1990.

Finsen, S. On moderation. In *Interpretation and Explanation in the Study of Animal Behavior.* Vol. 2: *Explanation, Evolution, and Adaptation,* M. Bekoff and D. Jamieson, eds. Boulder, Colo.: Westview Press, 1990.

Fitzgerald, M. Pain and analgesia in neonates. *Trends Neurosci.* 10: 344–346, 1987.

Francione, G. L. Animals, property and legal welfarism: "Unnecessary" suffering and the "humane" treatment of animals. *Rutgers Law Rev.* 46: 721–770, 1994.

Gallup, G. G. Chimpanzees: Self-recognition. *Science* 167: 86–87, 1970.

Gluck, J. P. Learning to see the animals again. In *Ethics in Practice: An Anthology,* H. LaFollette, ed. Cambridge, Mass.: Blackwell, 1997.

Goodall, J. *Through a Window: My Thirty Years with the Chimpanzees of Gombe.* Boston: Houghton Mifflin, 1990.

Hall, J. W. *Gone Wild.* New York: Delacorte Press, 1995.

Hiaasen, C. *Native Tongue.* New York: Fawcett Crest, 1991.

Holloway, M. Profile: Ruth Hubbard—Turning the inside out. *Sci. Am.* 272: 49–50, 1995.

Howard, W. E. An ecologist's view of animal rights. *Am. Biol. Teach.* 56: 202–205, 1994.

Jamieson, D. Experimenting on animals: A reconsideration. *Betw. Species* 1: 4–11, 1985.

Jamieson, D., and M. Bekoff. Carruthers on nonconscious experience. *Analysis* 52: 23–28, 1992.

Jamieson, D., and M. Bekoff. On aims and methods of cognitive ethology. *Phil. Sci. Assoc.* 2: 110–124, 1993.

Johnson, L. E. *A Morally Deep World: An Essay on Moral Significance and Environmental Ethics.* New York: Cambridge University Press, 1991.

Jolly, A. Conscious chimpanzees? A review of recent literature. In *Cognitive Ethology: The Minds of Other Animals. Essays in Honor of Donald R. Griffin,* C. A. Ristau, ed. Hillsdale, N.J.: Lawrence Erlbaum, 1991.

Kennedy, J. S. *The New Anthropomorphism.* New York: Cambridge University Press, 1992.

Kleiman, D. G., B. B. Beck, J. M. Dietz, L. A. Dietz, J. D. Ballou, and A. F. Coimbra-Filho. Conservation program for the golden lion tamarin: Captive research and management, ecological studies, educational strategies, and reintroduction. In *Primates: The Road to Sustaining Populations,* K. Benirschke, ed. New York: Springer-Verlag, 1986.

Kulli, J., and C. Koch. Does anesthesia cause loss of consciousness? *Trends Neurosci.* 14: 6–10, 1991.

Laurenson, M. K., and T. M. Caro. Monitoring the effects of non-trivial handling in free-living cheetahs. *Anim. Behav.* 47: 547–557, 1994.

Macdonald, D. *Running with the Fox.* New York: Facts on File, 1987.

Mann, M. D., D. A. Crouse, and E. D. Prentice. Appropriate animal numbers in biomedical research in light of animal welfare considerations. *Lab. Anim. Sci.* 41: 6–14, 1991.

Mason, W. A. Primatology and primate well-being. *Am. J. Primatol.* 22: 1–4, 1990.

Matthews, G. B. *Dialogues with Children.* Cambridge, Mass.: Harvard University Press, 1984.

Matthews, G. B. *The Philosophy of Childhood.* Cambridge, Mass.: Harvard University Press, 1994.

McGinn, C. Book review. *Times Lit. Suppl.* 1 August, 1991, p. 865.

McKibben, B. An explosion of green. *Atlantic Monthly* 275: 61–83, 1995.

McQuillan, K. *Elephants' Graveyard.* New York: Ballantine Books, 1993.

Midgley, M. *Animals and Why They Matter.* Athens: University of Georgia Press, 1983.

Morton, D. A fair press for animals. *New Scient.* 11 April, 1992, pp. 28–30.
Ng, Y.-K. Towards welfare biology: Evolutionary economics of animal consciousness and suffering. *Biol. Philos.* 10: 255–285, 1995.
Nicoll, C. S., and S. M. Russell. Analysis of animal rights literature reveals the underlying motives of the movement: Ammunition for counter offensive by scientists. *Endocrinology* 127: 985–999, 1990.
Orr, D. W. Professionalism and the human prospect. *Conserv. Biol.* 8: 9–11, 1994.
Perrigo, G., C. Bryant, L. Belvin, and F. S. Vom Saal. The use of live pups in humane, injury-free test for infanticidal behaviour in male mice. *Anim. Behav.* 37: 897–898, 1989.
Preston, R. *The Hot Zone.* New York: Random House, 1994.
Quinn, D. *Ishmael.* New York: Bantam/Turner, 1992.
Rachels, J. *Created from Animals: The Moral Implications of Darwinism.* New York: Oxford University Press, 1990.
Regan, T. *The Case for Animal Rights.* Berkeley: University of California Press, 1983.
Regan, T. Letter. *N.Y. Rev. Books* April 15, 1985, p. 57.
Report of the Laboratory Animal Society Working Party. The assessment and control of severity of scientific procedures on laboratory animals. *Lab. Anim.* 24: 97–130, 1990.
Ristau, C. A., ed. *Cognitive Ethology: The Minds of Other Animals. Essays in Honor of Donald R. Griffin.* Hillsdale, N.J.: Lawrence Erlbaum, 1991.
Rollin, B. E. *The Unheeded Cry: Animal Consciousness, Animal Pain and Science.* New York: Oxford University Press, 1989.
Russell, S. M., and C. S. Nicoll. A dissection of the chapter "Tools for Research" in Peter Singer's *Animal Liberation. Proc. Soc. Exp. Biol. Med.* 211: 109–138, 147–154, 1996.
Sebeok, T. A. A personal note. In *Man and Beast Revisited,* M. H. Robinson and L. Tiger, eds. Washington, D.C.: Smithsonian Institution Press, 1991.
Seligman, M. E. P., S. F. Maier, and J. H. Geer. Alleviation of learned helplessness in the dog. *J. Abnorm. Psychol.* 73: 256–262, 1968.
Shapiro, K., and P. B. Field. A new scale for invasiveness in animal experimentation. *Psychol. Ethical Treatment Anim.* 7: 1–4, 1987.
Shrader-Frechette, K. *Ethics of Scientific Research.* Lanham, Md.: Rowman and Littlefield, 1994.
Singer, P. Ten years of animal liberation. *N.Y. Rev. Books* January 17, 1985, p. 49.
Singer, P. *Animal Liberation,* ed. 2. New York: New York Review of Books, 1990.
Singer, P. Blind hostility: A response to Russell and Nicoll. *Proc. Soc. Exp. Biol. Med.* 211: 139–146, 1996.
Sprigge, T. L. *The Rational Foundation of Ethics.* New York: Routledge and Kegan Paul, 1988.
Stephens, M. L. *Maternal Deprivation Experiments in Psychology: A Critique of Animal Models.* Jenkintown, Pa.: American Antivivisection Society, 1986.
Taylor, P. W. *Respect for Nature: A Theory of Environmental Ethics.* Princeton, N.J.: Princeton University Press, 1986.
Tobias, M., ed. *Deep Ecology.* San Marcos, Calif.: Avant Books, 1988.
Tobias, M. *Rage and Reason.* New Delhi: Rupa, 1994.
Ulrich, R. *Rites of Life.* Kalamazoo, Mich.: Life Giving Experience, Inc., 1989.
Vincent, L. E., and M. Bekoff. Quantitative analyses of the ontogeny of predatory behaviour in coyotes, *Canis latrans. Anim. Behav.* 26: 225–231, 1978.
Westra, L. *An Environmental Proposal for Ethics: The Principle of Integrity.* Lanham, Md.: Rowman and Littlefield, 1994.
White, R. J. Letter. *Hastings Center Report.* November–December, 1990, p. 43.
Williams, B. *Morality: An Introduction to Ethics.* New York: Harper, 1972.

PART III

ASSESSING ANIMAL WELL-BEING

Asking the Animal

8

ANDREW N. ROWAN

The Search for Animal Well-Being

> Now it is impossible to point out the boundary line which separates the being that suffers from the being that does not suffer; and I defy anyone to establish any line of demarcation whatsoever between a being capable of pain and a being incapable of pain. . . .
>
> Now pain, taken in its profoundest sense, consists of two essential elements: a shock to the conscious self, the ego, in the first place; and, in the second place, the prolongation of the shock. If the self is not distinctly conscious, if it does not go so far as to assert itself by the separation of that self from the external world, we cannot say that pain is possible. No being suffers unless he is able to think that he suffers, and meditate on his suffering. . . .
>
> We have, therefore, the right to perform vivisection on beings which, because they possess no self hood, do not suffer. Now this absence of memory, consciousness, and intelligence extends assuredly over the whole of the vegetable kingdom, almost certainly over all the groups of the invertebrata, and also probably over all the inferior vertebrata.
>
> —C. Richet, *The Pros and Cons of Vivisection,* 1908

Animal well-being is an elusive concept. Most of us believe we can identify when an animal is in a state of distress or well-being, but when these states are examined more closely, it is clear that we do not have a detailed, operational grasp of what they entail. Well-being and distress are complex concepts that are difficult to define operationally. Also, in most discussions of well-being, the authors usually approach the concept by discussing the opposite—namely, states of pain, distress, and suffering (e.g., Mellor and Reid, 1994). This analysis will be no different. Although the title addresses animal well-being, I have approached the subject from the opposite end by examining animal distress and related concepts. If we can develop a better understanding of what these states might comprise, then we may begin to develop a better picture of animal well-being.

Pain and Distress

In the United States, the system of institutional animal care and use committees (IACUCs), established in the mid-1980s, is specifically charged with reducing the likely pain and distress that animals may experience when being used in research. Thus, IACUCs are required (by the U.S. Department of Agriculture regulators who oversee the conduct of animal research) to ensure that investigators have searched for alternatives if the research is likely to cause animal pain and distress, even if anesthetics and analgesics are used to prevent any pain and distress. By contrast, investigators do not have to demonstrate that they have considered or looked for alternatives if the animal research project is placed in the nonpainful category. This sends the implicit message that animal pain and distress is of greater public concern than animal death (euthanasia). Despite this regulatory emphasis on alleviating pain and distress, there is still relatively little systematic attention paid to such central issues as assessing pain and distress in animals and the potential impact of specific experimental procedures (e.g., dosing with chemical agents or infecting with pathogenic organisms) on animal well-being.

The systematic study of how to reduce animal pain and distress in research is not a trivial task. First, there is much conceptual confusion in the use of such terms as *pain, distress,* and *suffering*. Second, use of animals in the laboratory and the classroom varies widely. As Smith and Boyd (1991) state, "So wide is the variety of techniques used on animals in biomedical research, it is impossible to give general guidelines for reducing pain and distress." Third, animal pain, distress, and suffering are not easy to measure.

Aversive or distressing stimuli can take a variety of forms. Some cause physiological stress (e.g., injury, surgery, disease, starvation, and dehydration), some cause psychological stress (e.g., fear, anxiety, boredom, and lack of social interaction), some cause environmental stress (e.g. restraint, excessive noise, the presence of people or other species and chemicals), and some cause a mixture of stressors (ILAR, 1992). This chapter discusses some of the conceptual issues that have to be addressed in considering animal pain, distress, and suffering and also some of the difficulties of assessing the severity of such states.

The terms *pain, distress, anxiety, fear,* and *suffering* describe experiences and responses to experiences that are, in most cases, unpleasant and hence undesirable. Such terms are commonly used to describe both human and animal experiences, but what do we actually mean when we talk of pain, suffering, and distress? Dictionary definitions are circular and, hence, unhelpful. For example, in the 1967 unabridged *Random House Dictionary, pain* is defined as both a sensation of acute physical hurt or discomfort and emotional suffering and distress. *Suffering* is then defined as undergoing pain or distress. Obviously, this and other dictionaries view *pain, distress,* and *suffering* as synonyms, but closer analysis reveals they are not (Table 8.1).

Most discussion about animal pain and suffering concentrates on pain, not suf-

Table 8.1 Definitions of Pain and Distress Terms

Term	Definition	Illustration
Nociception	The process whereby potentially noxious and/or tissue-damaging stimuli cause special receptors (nociceptors to fire and send a nerve impulse along the nociceptive pathways.	Pain perception may occur, but only when such nerve impulses are processed in the central nervous system. For example, if one applies a heat stimulus to the foot of a high-level paraplegic whose lower spine is still functional, the paraplegic will withdraw the foot. No pain is felt, but the nociceptive reflex loop is still functional and will protect the foot from being burned.
Pain	An unpleasant sensory and emotional experience associated wtih actual or potential tissue damage or described in terms of such damage (IASP, 1979).	Pain terms are very variable. For example, people may talk of acute or chronic pain or sharp or dull pain. Pain is not physical or psychological, it is both.
Anxiety	An emotional state involving increased arousal and alertness prompted by an unknown danger that may be present in the immediate environment (Kitchen et al., 1987).	Unlike pain, anxiety is a diffuse sensation that has no specific location in the body. Unlike pain investigators, scientists who study anxiety have not developed a code of conduct to limit the extent of anxiety an animal may experience.
Fear	An emotional state involving increased arousal and alertness prompted by an experienced or known danger present in the immediate environment (Kitchen et al., 1987).	
Distress	A state in which the organism is unable to adapt to an altered external or internal environment.	In acute distress, the organism will try to escape, but in chronic situations its response may be maladaptive (eg., learned helplessness).
Suffering	A highly unpleasant emotional response usually associated with pain and/or distress (Kitchen et al., 1987).	The adjective "emotional" stresses the affective nature of suffering. Suffering involves a threat to the "personhood" or self-concept rather than simply the organic body of a human individual.

fering. A fairly recent report from the Netherlands, *Definitions of Pain, Stress, and Suffering and the Use of These Concepts in Legislation on Animal Suffering,* includes almost no discussion of suffering itself, although the term comes up frequently in the text (Voorzanger and de Cock Buning, 1988). Similarly, the report on animal pain and distress by the Institute of Laboratory Animal Resources (ILAR) (1992) defines and discusses pain, distress, anxiety, fear, and discomfort but deliberately excludes any discussion of animal suffering. For the ILAR working party, suffering could not be defined operationally and therefore could not be reliably assessed. Like obscenity, people are confident they can recognize suffering when they see it, but they cannot define what it is. Pain is also very complex and, ultimately, a private phenomenon, but it is nevertheless considered easier to measure and to ground in the empirical world of biomedical research.

Identifying Pain In Animals

Although most humans can report whether they feel pain, animals cannot, which has led to problems in acknowledging animal pain (see Beynen et al., 1987, and Phillips, 1993, for examples). Nonverbal human infants were, until recently, also denied the capacity to fully experience pain, which confirms the importance of verbal report in legitimating pain perception (Anand and Hickey, 1987). Nevertheless, it is possible to study pain perception in animals (just as it is in human infants).

Typically, we resort to studies of nonverbal behavior, such as moaning and crying, writhing, and wriggling, to infer the presence of pain perceptions in animals. Pain also has typical physiological and neurophysiological correlates, which, unlike the phenomenological (or felt!) occurrence that is pain, are subject to direct empirical investigation. For example, nociceptors (the nerve endings that, when stimulated, are associated with pain perceptions) have been found in all mammals and in other vertebrates. In addition, direct, percutaneous recordings in human subjects have demonstrated that feelings of pain are correlated with activity in the small myelinated and unmyelinated nerves. Research on anesthetized animals indicates that these same nerve fibers are activated exclusively (or most potently) by stimuli of noxious intensity. Such nerve fibers appear to be present in all vertebrates. Similarities between humans and animals have also been demonstrated in the central nervous system (CNS) pathways involved in pain perception.

Knowledge about the neurochemistry of pain is surging ahead, beginning with the discovery of the body's natural opiates—the endorphins. Comparative studies indicate that endorphins and other chemicals associated with painful stimuli (e.g., substance P) are widespread throughout the animal kingdom (even earthworms have endorphins). But endorphins are now known to have a variety of roles and are not limited to pain pathways. Therefore, it would be dangerous to extrapolate simply from neurochemical similarities and infer that earthworms experience pain as humans do. Humans with high spinal cord injuries may have intact lower spinal reflexes and spinal neurochemistry that would result in withdrawing a foot from a noxious heat stimulus even though the brain perceives no pain.

Two pain threshold levels have been described in humans. The first is the pain detection threshold, which is defined as the smallest stimulus that is felt as painful 50 percent of the time. A similar threshold can be measured in animals and is identified by the appearance of escape behavior. The pain detection threshold level seems to be the same in both humans and other vertebrates. The second level is the pain tolerance threshold. Under similar conditions, primates and humans show the same pain tolerance threshold, but this threshold is very plastic and subject to wide variation, depending on the environmental conditions and the psychological state of the human or animal.

Thus, reasoning from analogy, neuroanatomy, neurophysiology, neurochemistry, and behavioral observations, most people conclude that animals—or at the very least the warm-blooded vertebrates—probably experience pain that is qualitatively and quantitatively similar to that experienced by humans. Nevertheless, there are still many scientific and metaphysical questions regarding human and animal pain perception.

Puzzles and Issues: Pain

Pain is a phenomenological occurrence that requires some conscious experience in the individual who is experiencing the pain. Thus pain, in ordinary parlance, is, by definition, felt. From time to time, individuals may refer to unconscious pain (e.g., Carruthers, 1992) and justify such terminology by referring to a variety of behaviors that seem to indicate "unconscious" pain. For example, we move our limbs out of potentially painful conformations when we are asleep (people who are congenitally insensitive to pain do not do this). However, such unconscious behavior could be stimulated by the sort of simple reflex nociceptive nerve networks mentioned. Therefore, we do not require use of such terminology as *unconscious pain.*

Because pain involves something's hurting, pain is necessarily negative or unpleasant in character. That is not to say someone may not be willing to tolerate pain or even in some sense enjoy it. A person puts up with great pain in the dentist's chair for the sake of dental health. We may "enjoy" the pain resulting from yesterday's exercise because we know the workout that caused it was good for us. But for pain to be a net positive experience, something valuable to the subject must override pain's unpleasantness or its prima facie negative character (Pitcher, 1990). Again, if it does not hurt, it cannot be defined as pain—and hurting is a prima facie negative experience for the individual who hurts.

Lobotomized Individuals

Several interesting cases challenge the commonsense notion that pain is prima facie unpleasant. For example, patients who have undergone prefrontal lobotomies claim that they experience pain but do not mind it (Damasio, 1994). They retain the startle reflex (they jump when pricked), indicating that their nociceptive pathways are functioning. However, they can watch a needle being pushed into their flesh with little or no sign of concern. These patients have been reported to say that they feel the "little" pain but not the agony or "big" pain (Melzack, 1973).

Is it possible, though, to have pain without minding it (at all)? Lobotomized individuals report feeling something—which they call pain—so either there is no pain and they are misidentifying pressure sensations and nociception, or there is pain. We tend to take them at their word when they say they have pain, but there are at least two alternative explanations for their odd behavior. First, their pain may be perceived as mild (to make sense of their lack, or relative lack, of concern about it). There are certainly grades of pain, and we have all experienced pain that is perfectly tolerable, at least in the short term. However, having a sharp object pushed into a person's leg does not seem to be consistent with the idea of mild pain. Second, the dulled perception of pain may be related to the removal of the prefrontal cerebral cortex, which has the effect of producing an individual with very flat emotional responses. As argued later, suffering pain involves an emotional reaction, and if a person's emotional reactions have been virtually eliminated, then so may the suffering. Damasio (1994) provides some compelling examples of the importance of emotional reactivity in producing suffering.

More than 30 years ago, Brain (1963) suggested that animals, with their smaller prefrontal lobes, could be similarly unconcerned by pain while still displaying the nociceptive startle reflex. This idea is a reincarnation of the Cartesian view of animals as unfeeling machines developed 400 years ago. This view keeps reappearing, and Carruthers (1992) presents a modern version of the argument. However, experimental evidence, especially regarding the effectiveness of analgesic compounds that eliminate pain in humans in also eliminating escape and other pain behavior in animals, make such Cartesian arguments increasingly unlikely.

Masochism: Enjoying Pain?

Some of Pavlov's experiments illustrate another difficult case in trying to develop a coherent and consistent phenomenology of pain. Pavlov reported that, in one series of studies, dogs wagged their tails and displayed other signs of pleasure while receiving electric shocks that they associated with subsequent food rewards (cited in Pitcher, 1970, p. 483). Without implying equivalence, sadomasochists demonstrate the same sort of behavior and apparently enjoy their pain. However, in both cases, the pain could still be unpleasant, but the dog or sadomasochist has in some way become conditioned to focus on the anticipated reward rather than reacting to the aversive stimulus as most "normal" animals and people do. The overall situation sets up a conflict that has to be resolved, and, in these examples, anticipation wins.

Benefits of Pain

Although pain is unpleasant in character, it is at the same time good, in the sense of advantageous, from an evolutionary perspective. Patrick Bateson (1991) of Cambridge University suggests that pain may be advantageous in at least these ways:

1. It allows the animal to distinguish at the peripheral level between potentially dangerous stimulation and intense but harmless stimulation that carries potentially useful information.
2. It enables an animal to learn to avoid conditions previously associated with potentially harmful stimulation.
3. It enables an animal to give top priority to escaping from or removing potentially harmful stimulation and to avoiding conditions previously associated with potentially harmful stimulation.

To illustrate the devastating consequences of being unable to feel pain, we have only to point to those rare and unfortunate individuals who are congenitally insensitive to pain. Children with this syndrome do not learn to avoid danger (e.g., open stairs, hot stoves), and most individuals do not survive beyond their early twenties, by which time they are horribly scarred.

Interestingly, one case study reported that the affected individual did not perceive pain but was, nevertheless, terrified of surgery. This paradigmatic example of suffering in the absence of pain (Thrush, 1973) can be explained by noting that the surgery threatened the psychological personhood of this individual without in any way causing pain. Nonetheless, it is not absolutely necessary for a person to perceive pain to be protected from injury. As mentioned earlier, a functioning nociceptive reflex loop can protect an organism from at least some potentially damaging stimuli, specifically acute noxious insult (see discussion of pain and suffering in insects).

Invertebrate Pain?

Although most people assume that vertebrates perceive pain, the situation is not as clear for most invertebrates (e.g., Eisemann et al., 1984, and Fiorito, 1986, discuss the problem of pain in insects). However, the common octopus, with its large CNS and complex behaviors (e.g., Smith and Boyd, 1991) has been given the benefit of the doubt in Great Britain and is now protected under the Animals (Scientific Procedures) Act of 1986.

Some argue that insects do not perceive pain but that it is difficult to be certain. Others are not sure but believe that insects should be accorded the benefit of the doubt (e.g., Wigglesworth, 1980). Still others (e.g., Eisemann et al., 1984; Fiorito, 1986) argue that insects do not perceive pain, although they might avoid some aversive stimuli. The conclusion that insects do not perceive pain is based on several lines of reasoning.

First, although insects have complex nervous systems, they lack the well-developed central processing mechanisms found in mammals and other vertebrates (and the octopus) that appear to be necessary to feel (perceive) pain. Second, insects apparently do not have a nerve fiber system equivalent to the nociceptive fibers found in mammals, which, however, does not mean that they do not have some nerve fibers that carry nociceptive signals. Third, the behavior of insects that are faced with noxious stimuli can usually be explained as a startle or nociceptive protective reflex. In some cases (e.g., locusts being eaten by fellow locusts), insects display no signs that the tissue damage that is occurring is aversive.

The conclusion that insects do not perceive pain appears to contradict the claim that pain confers important survival advantages. However, simple nociceptor reflex loops (producing the startle reflex) that involve no pain perception could confer sufficient evolutionary advantage in short-lived animals (like insects) that rely on a survival strategy that involves the production of very large numbers of individuals.

Suffering

Suffering is a widely used and abused colloquial term that has been subjected to very little careful analysis, even in the case of human suffering. Cassell (1982), one of the few to address the biological and psychological roots of human suffering, argues that suffering occurs when the integrity of a person (not the body) is perceived to be compromised or threatened in some way. (Personhood is defined in terms of an individual's mental life and is distinguished from the organic body). Damage to organic tissues can (and often does) lead to suffering, but for Cassell, the psychological reaction to such damage is the key to understanding the idea of suffering.

The notion that suffering arises from a perceived threat to the integrity of a person has significant ramifications for any discussion of animal suffering. Animals would, according to this definition, suffer only if they possess to some degree the qualities of personhood. In a later analysis, in which he specifically addresses the issue of animal suffering, Cassell (1989) argues that only beings with a sense of the future and a sense of self are capable of experiencing suffering. Some animals do appear to have a sense of self (e.g., chimpanzees) and a sense of the future—or, at least, seem to be able to anticipate future events. How far such abilities extend through the animal kingdom would necessitate a much more detailed analysis than is possible here. We could also argue that only animals that are capable of affective (e.g., emotional) responses might be included in the category of beings capable of suffering (e.g., Damasio, 1994).

It is quite clear that few, if any, people use suffering in the narrower sense articulated here, and even scientists who refuse to use the term *suffering* in referring to animal distress (ILAR, 1992) nevertheless argue vehemently that animals (including invertebrates) are capable of suffering. However, the colloquial term *suffering* has such broad meaning (see dictionary definitions) that it cannot be used profitably (even after careful definition) to assess the severity of aversive stimuli to animals or even to discuss the level of distress experienced by animals. In this chapter, I will not dwell further on whether animals experience suffering. Instead, I wish to examine one fairly specific element of affective machinery—namely, an animal's capacity to experience anxiety.

Anxiety and Its Role in Distress

Considerable attention has been paid to the use of appropriate levels of painful stimuli by researchers who study pain in animals. For example, the International Association for the Study of Pain has set up guidelines (IASP, 1979) in which researchers are urged to design only projects in which animals can terminate any painful stimulus and thus control the level of pain they experience. Some of the simpler pain research protocols (e.g., the tail flick and hot plate tests) involve systems where the animal chooses to end the pain by moving itself or its tail away from the stimulus. The tail flick and hot plate devices are also fitted with automatic cutoff switches so that, in the event the analgesia under study is very effective, the animal will still not suffer any tissue damage.

Similarly, we can develop systems that allow animals to "volunteer" for pain research by offering them a highly desired food or drink. Such animals are willing to accept some painful stimuli to gain the reward. However, at the pain tolerance threshold, they voluntarily choose not to participate any further. (Primates appear to have tolerance thresholds very similar to those of humans.) In all of these cases, the research protocol allows the animal to control when the painful stimulus is terminated. There are some studies (e.g., of chronic pain) in which such refinements are not possible, but even here, pain researchers have tried to ensure that the animals do not endure a significant level of pain (Casey and Dubner, 1989).

By contrast, researchers who study anxiety in animals, which is arguably just as distressing to animals (if not more so), have not developed similar guidelines and approaches. Some may not have paid attention to animal anxiety because they do not believe that animals can be anxious. For example, Cassano (1983), a psychiatrist, has stated that "fear is a primitive state of mind found throughout the animal kingdom, whereas anxiety is part of conscious experience and takes shape as a typically human function or attitude. Thus, the age of anxiety could be said to begin with the emergence of Homo sapiens" (p. 288).

What the difference might be between fear and anxiety is not exactly clear. A person might fear some definable danger, whereas anxiety may refer to that state of uneasiness where the threat is undefined and elusive. Cassano does not provide any clear distinguishing characteristics between fear and anxiety. However, there is at least one relatively clearly defined neural substrate that appears to be involved in mediating anxious states, and this substrate, interestingly, was found to be present in all vertebrates but in none of the invertebrates examined (Nielsen, Braestrup, and Squires, 1978). This substrate has come to be known as the benzodiazepine receptor (because it binds the anxiolytic benzodiazepine drugs such as diazepam [Valium] with high affinity), but it also binds alcohol and the barbiturate drugs, which also diminish feelings of anxiety.

Building on investigations of drug binding to the benzodiazepine receptor and its subsequent behavioral effects, Gray (1982) has produced a comprehensive theory of anxiety in which he argues that "'human anxiety,' or something very like it, exists also in animals" (p. 17). Gray recognizes that many people may find this conclusion hard to accept because of the common belief that anxiety is an almost uniquely human state, dependent on such complex cognitive capacities as the ability to foresee the future, to form a self-image, or to imagine one's own mortality. Nevertheless, he argues that the observed effects of such antianxiety drugs as alcohol, the barbiturates, and the benzodiazepines in animals are so similar to the observed effects of these drugs in humans that it seems more parsimonious (following Morgan's canon, Morgan, 1894) to argue that these agents act upon a state in animals that is similar to the human state of anxiety.

Some of the animal models used to discover and study anxiolytic agents clearly cause some animal distress. One such model, the Vogel antianxiety test, sets up a conflict situation in which a water-deprived animal is given access to water but also receives an electric shock on a variable schedule when it drinks. This situation produces an anxious response in the animal and depresses the drinking frequency and amount of water consumed. Under the influence of anxiolytic (or antianxiety) agents, an animal will drink more fluid and drink frequently.

Another animal model of anxiety is the social interaction test (File, 1987), which measures the amount of time two male rats spend socializing in an enclosure where both the familiarity and the level of lighting can be manipulated. Undrugged rats show the highest level of social interaction when the enclosure is familiar and under low light. Social interaction declines if the enclosure is unfamiliar or brightly lit. Anxiolytic drugs prevent this decline. This social interaction test approximates the natural situation much more closely than the Vogel test and is, intuitively, more humane because the animal is allowed greater control of the level of stress it experiences.

Research has also identified compounds that cause anxiety and that bind to the benzodiazepine receptor in the CNS. The best known of these are the beta-carbolines, which when administered to humans cause intense inner strain and excitation, increased blood pressure and pulse, restlessness, increased stress hormone levels in the blood, and stereotyped rocking motions. One volunteer experienced such severe anxiety that he had to be physically restrained and injected with a benzodiazepine, which provided relief within five minutes (Dorow et al., 1983). The administration of beta-carbolines to primates caused piloerection, struggling in the restraint chair, increased blood pressure and pulse, increased stress hormone levels in the blood, and increased vocalization and urination (Ninan et al., 1983).

The similar reactions of human volunteers and primates to the beta-carbolines does not prove that both humans and primates experience the same sort of anxiety, but it is hard to argue that animal anxiety is not a significant cause of animal distress and suffering. Gray (1982) has suggested that anxiety may have evolved from a biological behavioral system—the behavioral inhibition system (BIS)—that may confer an evolutionary advantage on those animals that have it. The BIS stimulates a state of alertness to novelty in an animal's environment and makes the animal less likely to rush into danger. Excessive stimulation of the BIS can clearly cause animal distress and suffering, as Reese's (1979) studies of a strain of anxious pointer dogs illustrates.

Reese's inbred strain of purebred pointers produce litters in which about 80 percent of each litter is distinctly pathological. The prominent characteristics of the nervous animals compared with their normal counterparts include excessive timidity, hyperstartle responses, reduced exploratory activity, and frequent rigid immobility in the presence of humans. Reese (1979) notes that the laboratory housing is not a dog's paradise (although they received the Lassie Award for excellent treatment of dogs) but that the various negative experiences cannot explain the nervous behavior because the normal dogs remain friendly and cooperative. In addition, the nervous dogs appear to interact with other dogs normally and display relatively normal behaviors with human handlers when treated with anxiolytic drugs.

The Complexity of Anxiety

Although the distribution of the benzodiazepine receptor in vertebrates appears to provide a relatively "clean" distinction between sentient vertebrates and nonsentient invertebrates, research over the past decade has produced a host of confounding factors. First, other benzodiazepine binding sites have now been demonstrated in invertebrates (Lummis, 1990). These receptors are found in nonnervous tissue, and they are different from those found in the CNS of vertebrates. Nevertheless, the presence of these different benzodiazepine binding sites confuses the speculative idea that the presence of benzodiazepine receptors is indicative of sentience.

Second, other receptors that mediate anxiety and other anxiolytic drugs have now been identified. For example, cholecystokinin peptides and their receptors appear to be involved in mediating anxiety and panic (Derrien et al., 1994). Handley and McBlane (1993) describe a number of drugs (including the increasingly popular anxiolytic, buspirone) that act through 5-hydroxytryptamine (5-HT; serotonin) receptors to mediate anxiety in both humans and animals. Thus, anxiety cannot be attributed to a single neurochemical system in the CNS. Nevertheless, it is abundantly clear from the pharmacology of anxiety in both animals and humans that anxiety can be a significant cause of distress and suffering in animals.

Conclusions

What do these findings and speculations about animal pain, suffering, and anxiety tell us about well-being? First, we have to broaden our concerns about pain to include a number of other states, such as anxiety, that are capable of producing considerable suffering. Second, as suffering is conceived in the discussion in this chapter, it may not be distributed as widely through the animal kingdom as the vernacular use of the term might suggest. For example, Damasio (1994) argues that suffering arose in creatures that have rather sophisticated neural machinery capable of large-scale storage of a multitude of categories of objects and events and capable of manipulating those representations and fashioning new creations by means of novel combinations.

This analysis does not deal at all with the promotion of well-being. Presumably, we have some responsibility not only to minimize animal pain, distress, and suffering but also to enrich and enhance the existence of animals that we use and keep for human benefit. This point is what seems to lie behind efforts to develop environmental enrichment programs for zoo and laboratory animals and the pressure to change minimum standards of animal care into optimal standards. However, if our understanding of animal pain, distress, and suffering is confused and incomplete, our knowledge of what might constitute animal well-being is in even worse shape. What do we have to do to promote an animal's well-being?

Presumably, rewarding and reinforcing stimuli and environments are necessary to promote well-being, but it is not quite that simple. A dog may find chocolate very rewarding, but we know it is not good for the animal. If an animal is provided with food and water in a safe environment, why do we not consider this care necessarily sufficient to maintain a state of well-being? Experiments indicate animals preferentially work for food rather than eat what is freely available, which indicates that foraging activity is in itself rewarding. What about the topic of play? Play behavior is intuitively associated with pleasure and well-being, but is this a safe assumption? If play is an indicator of well-being, should we try to increase its frequency, even among adult animals? If so, how would we increase the incidence of such behavior? (In fact, while we may think we know play when we see it, behavioral scientists argue endlessly over how to define play and what such behavior might mean; see Mitchell, 1990).

For the moment, we are left with far more questions than answers. Fortunately, there has been an increase in attention to these issues by research scientists and other academics. The result is more experimental data that address animal distress and well-being issues and more debate about the conceptual issues. I hope that all this activity will lead to improvements for both animals and the humans who rely on or appreciate them.

References

Anand, K. J. S., and P. R. Hickey. Pain and its effects in the human neonate. *N. Engl. J. Med.* 17: 1321–1329, 1987.

Bateson, P. The assessment of pain in animals. *Anim. Behav.* 42: 827–839, 1991.

Beynen, A. C., V. Baumans, A. P. M. G. Bertens, R. Havenaar, A. P. M. Hesp, and L. F. M. van Zutphen. Assessment of discomfort in gallstone-bearing mice: A practical example of the problems encountered in an attempt to recognize discomfort in laboratory animals. *Lab. Anim.* 21: 35–44, 1987.

Brain, L. Animals and pain. *New Scientist* 18: 380–381, 1963.

Carruthers, P. *The Animals Issue.* Cambridge: Cambridge University Press, 1992.

Casey, K. L., and R. Dubner. Animal models of chronic pain: scientific and ethical issues. *Pain* 38: 249–252, 1989.

Cassano, G. B. What is pathological anxiety and what is not? In *The Benzodiazepines: From Molecular Biology to Clinical Practice*, E. Costa, ed. New York: Raven Press, 1983.

Cassell, E. J. The nature of suffering and the goals of medicine. *N. Engl. J. Med.* 306: 639–645, 1982.

Cassell, E. J. What is suffering? In *Science and Animals: Addressing Contemporary Issues,* H. N. Guttman, J. A. Mench, and R. C. Simmonds, eds. Bethesda, Md.: Scientists Center for Animal Welfare, 1989.

Damasio, A. R. *Descartes' Error: Emotion, Reason and the Human Brain.* New York: G. P. Putnam's Sons, 1994.

Derrien, M., I. McCort-Tranchepain, B. Ducos, B. P. Roques, and C. Durieux. Heterogeneity of CCK-B receptors involved in animal models of anxiety. *Pharmacol. Biochem. Behav.* 49: 133–141, 1994.

Dorow, R., R. Horowshi, G. Paschelke, M. Amin, and C. Braestrup. Severe anxiety induced by FG7142; a beta-carboline ligand for benzodiazepine receptors. *Lancet* 2: 98–99, 1983.

Eisemann, C. H., W. K. Jorgensen, D. J. Merrit, M. J. Rice., B. W. Cribb, P. D. Webb, and M. P. Zalucki. Do insects feel pain? A biological view. *Experientia* 40: 164–167, 1984.

File, S. E. The search for novel anxiolytics. *Trends Neurochem. Sci.* 10: 461–463, 1987.

Fiorito, G. Is there pain in invertebrates? *Behav. Proc.* 12: 383–386, 1986.

Gray, J. A. *The Neuropsychology of Anxiety*. New York: Oxford University Press, 1982.

Handley, S. L., and J. W. McBlane. 5HT drugs in animal models of anxiety. *Psychopharmacology* 112: 13–20, 1993.

IASP (International Association for the Study of Pain). Report of International Association for the Study of Pain, subcommittee on taxonomy. *Pain* 6: 249–252, 1979.

ILAR (Institute of Laboratory Animal Resources). *Recognition and Alleviation of Pain and Distress in Laboratory Animals*. Washington, D.C.: National Academy of Sciences, 1992.

Kitchen, H., A. L. Aronson, J. L. Bittle, C. W. McPherson, D. B. Morton, S. P. Pakes, B. E. Rollin, A. N. Rowan, J. A. Sechzer, J. E. Vanderlip, J. A. Will, A. S. Clark, and J. S. Gloyd. Panel report on the colloquium on recognition and alleviation of animal pain and distress. *J. Am. Vet. Med. Assoc.* 191: 1186–1191, 1987.

Lummis, S. C. R. GABA receptors in insects. *Comp. Biochem. Physiol. C* 95: 1–8, 1990.

Mellor, D. J., and C. S. W. Reid. Concepts of animal well-being and predicting the impact of procedures on experimental animals. In *Improving the Well-Being of Animals in the Research Environment,* R. M. Baker, D. Jenkin, and D. J. Mellor, eds. Glen Osmond, South Australia: ANZCCART, 1994.

Melzack, R. *The Puzzle of Pain*. London: Penguin, 1973.

Mitchell, R. W. A theory of play. In *Interpretation and Explanation in the Study of Animal Behavior,* Vol. 1, M. Bekoff and D. Jamieson, eds. Boulder, Colo.: Westview Press, 1990.

Morgan, C. L. *An Introduction to Comparative Psychology.* London: Walter Scott, 1894.

Nielsen, M., C. Braestrup, and R. F. Squires. Evidence for a late evolutionary appearance of a brain-specific benzodiazepine receptor. *Brain Res.* 141: 342–346, 1978.

Ninan, P. T., T. M. Insel, R. M. Cohen, J. M. Cook, P. Skolnick, and S. K. Paul. Benzodiazepine receptor-mediated experimental anxiety in primates. *Science* 218: 1332–1334, 1983.

Phillips, M. T. Savages, drunks and lab animals. *Society and Animals* 1: 61–82, 1993.

Pitcher, G. The awfulness of pain. *J. Philosophy* 68: 481–492, 1970.

Reese, W. C. A dog model for human psychopathology. *Am. J. Psychiat.* 136: 1168–1172, 1979.

Richet, C. *The Pros and Cons of Vivisection.* London: Duckworth, 1908.

Smith, J. A., and K. M. Boyd. *Lives in the Balance. The Ethics of Using Animals in Biomedical Research.* Oxford: Oxford University Press, 1991.

Thrush, D. C. Congenital insensitivity to pain: A clinical, genetic and neurophysiological study of four children from the same family. *Brain* 96: 369–386, 1973.

Voorzanger, B., and T. de Cock Buning. *The Definitions of Pain, Stress, and Suffering, and the Use of These Concepts in Legislation on Animal Experiments.* Proefdier en Wetenschap No. 1. Leiden, the Netherlands: Leiden University, 1988.

Wigglesworth, V. B. Do insects feel pain? *Antenna* 4: 8–9, 1980.

Young, J. Z. The organization of a memory system. *Proc. R. Soc., Ser. B* 164: 285–320, 1965.

Young, J. Z. *The Anatomy of the Nervous System of* Octopus vulgaris. Oxford: Clarendon Press, 1971.

9

MARIAN STAMP DAWKINS

Considering Animal Welfare from the Animal's Point of View

I would like to argue that an essential component of animal welfare is that its conditions and way of life are satisfactory from the animal's point of view. Without this component, the study of animal welfare would consist solely of the study of animal health and would be no more problematical than discovering the circumstances that favored their preservation and a long life. We could, by direct analogy, talk about the welfare of plants or, for that matter, the welfare of oil paintings. But even though we may put considerable effort into caring for plants and precious human artifacts, there is an important reason why we talk about their "protection" or "conservation" rather than their "welfare." To use the words *welfare* or *well-being* as we do with animals implies that we believe we are dealing with conscious beings capable of experiencing that well-being or its opposite, suffering.

We therefore cannot seriously study animal welfare without in the end tackling the problem of how to study conscious experiences in species other than our own (Dawkins, 1990). However, in the context of animal welfare, I shall argue that we should be less concerned with what have been called "cognitive" processes in animals and more concerned with the neglected areas of their emotional experiences and their motivation to perform different behaviors.

In ourselves, the ability to experience emotions does not require great feats of intellect. Even though Damasio (1994) argued that emotion and reason are very closely linked in our brains, it is nevertheless true that you do not need to be particularly clever to feel pain or to experience hunger. Consequently, I believe that the study of animal welfare should focus primarily on animal emotions and not necessarily on their intellect. The clever things that some animals can do—their cognitive abilities to think and to reason are important—are certainly relevant to animal welfare, but their relevance is primarily in making people who might otherwise dismiss all nonhumans as stupid stop and consider the possibility that their welfare should be taken seriously. Without really knowing why, many people believe that having a mind is key to being given moral consideration.

The really important issue in animal welfare, as Jeremy Bentham pointed out two centuries ago, is not thinking or using language, however, but suffering. Animal welfare studies should thus take notice of animal intellect and welcome recent scientific interest in animal minds (Dawkins, 1993; Gould and Gould, 1991; Griffin, 1992) but nevertheless make their own primary focus that of animal emotions, particularly the negative emotions we call suffering. I shall argue that, difficult though that is, this focus is now possible and that the best way we have is through the study of certain kinds of behavior. Although there are many behavioral measures we can use, which, indeed, have been used successfully (Broom and Johnson, 1993), I wish to concentrate on one particular approach, one that uses the animal's point of view as an indicator of what kinds of emotions the animal is experiencing.

The focus of this chapter is *suffering,* which is difficult enough to define, rather than on the more positive term *welfare,* which leads us into a veritable minefield of definitions even with our own species (try giving a single definition of *human happiness* or *human welfare*). One day, of course, we hope that the theory (and practice) of animal welfare will have progressed to the point where we can talk about positive welfare. But for the moment, the negative side—suffering—still forms so large a problem that I make no apology for concentrating on the absence of poor welfare rather than (yet) achieving presence of positive welfare.

Suffering, as we use it to apply to human beings, covers a very wide range of different states. We talk about suffering from fear, suffering from thirst, suffering from an illness, suffering from grief, suffering from boredom, and suffering from overwork, to name just a few. What all these states have in common is that (1) they are all unpleasant and are consciously experienced as such, (2) they are acute, persistent, or both (we would not call a mild itch suffering, but a violent or persistent itch might well cause suffering), and (3) they are all states that we would rather not be in and that, if it is in our power, we attempt to alleviate or get out of.

The study of animal suffering is, by analogy, the study of a wide range of states such as pain, fear, boredom, and frustration. Our problem is to recognize these states in species whose behavior and external appearance may be very different from our own. The first thing to realize is that there are no easy answers. No simple litmus test can provide for an instantaneous answer. What we have to do is collect different sorts of evidence that, together, can bring us toward an answer (Dawkins, 1980).

We should look for three major sorts of evidence: (1) physical ill health and injury, (2) physiology, and (3) behavior. I have summarized these types and the uses to which they have been put in Table 9.1; although I wish to concentrate here on one of the behavioral measures, I certainly do not wish to imply that this is the only way of studying suffering. On the contrary, if all we had was the animal's point of view, we would be prone to serious misjudgments about animal welfare in just the same way that, if all we had was a child's point of view about going to the dentist, we might make serious errors about what was best for the child's welfare in the long run. The animal's point of view should never be considered in isolation from other measures of welfare and suffering that are also available to us. The study of animal emotions is too difficult a pursuit for us to ignore any of the evidence that may be available to us.

Table 9.1 Sources of evidence of suffering in nonhuman animals

1. Ill health, overt signs of disease Injuries Impaired growth Impaired reproduction Reduced life expectancy
2. Physiological measures Elevated heart rate Elevated glucocorticoid levels Suppressed immune system responses
3. Behavior stereotypies and other abnormal behavior Rebound effects Measurements of motivation, especially the animals' point of view

Note: For more details, see Dawkins (1980) and Broom and Johnson (1993).

The states we call suffering are so diverse both in what causes them and in what they feel like when we experience them that we might wonder how it is that we can apply the one word *suffering* to them. As I have already mentioned, they do have in common that they are all very unpleasant and—something that is crucial to our attempts to study animal suffering—they are all states that we do our best to get out of if we possibly can. So if we are suffering from thirst or from fear, we have private mental experiences that other people can know nothing about, but we also do things that other people can quite easily observe. Suffering from thirst, for example, is accompanied by searching for water, possibly being prepared to pay a lot of money for something to drink, and then drinking a great deal when we finally are able to. Our behavior, in other words, is an overtly observable manifestation of our private subjective experience of thirst. I am going to show that a similar line of argument can be taken with other species and used to give important information about the circumstances in which they suffer.

The first point to be made is that the capacity to suffer is not arbitrary. It is not a useless "extra" added on to organisms to make their lives difficult. On the contrary, it is as much part of their equipment for survival as the color of their fur or the shape of their feathers. For most biologists, Darwin's theory of natural selection provides the most plausible explanation we have for the evolution of animal bodies and of the particular forms and shapes they take. A great deal of work now supports the idea that an animal's behavior, too, is the result of natural selection (e.g., Krebs and Davies, 1993). Animals that behaved in some ways, such as staying close to other members of their species or building a protective nest for their offspring, survived and reproduced more successfully than those that stayed on their own or left their young in places open to predation. Their offspring survived and, in turn, passed the favorable behavior on to the next generation. The most fruitful way to see the capacity to suffer is as yet another part of this "equipment for survival." Animals that suffered from fear and, through experiencing this unpleasant state, removed themselves most effectively from a source of danger were less likely to be eaten by a predator than those that stayed in the danger zone. Animals that suffered from thirst and made a great effort to find water were less likely to die of dehydration than those that carried on with other activities and failed to find water. Suffering is, then, an early warning system for death and destruction, an indication that, if nothing is done, the organism or at least the organism's reproduction will be in peril (Dawkins, 1990).

The phrase *if nothing is done* is the key to why we should be particularly concerned with the suffering of animals in captivity. Wild animals are constantly faced with threats to their survival—food shortages, hungry predators, intense heat or cold. In this sense, wild animals suffer, too. But whereas a wild animal that experiences fear of a predator can take steps to remove itself from the source of danger by fleeing over the horizon, an animal confined in a cage and exposed to what it takes to be a predator will attempt to flee yet cannot do so. Even if it is highly motivated to escape, its cage prevents it. It is left with an alarm system permanently turned on, an emergency drill activated but prevented from being carried out. The suffering that, in nature, has a high chance of ensuring its survival remains because the behavior that would remove the source of danger and the source of suffering cannot be carried out. So the difference between wild and captive animals is not that wild animals do not suffer (because suffering in its rightful place is an aid to survival in the wild) but that captive animals are liable to remain longer in the state we call suffering. They cannot turn it off because the behavior that would do so in the wild is prevented.

This difference raises an important point about the relationship between what is adaptive or good for the animal's survival, on the one hand, and what animals actually do, on the other. The theory of natural selection, while it leads us to see most animal structures and behavior as in some way helpful to survival, does not lead us to expect that absolutely everything that animals do will be in their best interests (e.g., Dawkins, 1982). One of the major reasons why animals may not always do what is best for them is that many of them are kept in conditions that are very different from the ones in which they evolved.

For example, imagine a hypothetical species of migratory bird that normally flies long distances each autumn and spring. In the autumn, the birds fly south and escape northern winters, and in the spring they move hundreds of miles north to breed. Their migratory behavior is adaptive in the sense that any individuals that do not migrate have no chance of surviving or breeding, even though the flights are hazardous and many individuals will not survive the journeys. Now imagine that some members of this species are captured and kept in large flight cages so that they cannot migrate. They are well cared for and protected from both the elements and predators. In fact, their chances of survival over the winter are considerably greater than those of their wild counterparts that migrate in the normal way. However, the captive birds show intense "migratory restlessness" at the time of year when wild members of their species are migrating. They attempt to escape from their cages in the direction that they would normally be migrating and flutter repeatedly at the walls of the cage. In reality, their best chance of surviving to the following spring is to stay in the aviaries. However, what they are most motivated to do is to fly because nothing in their evolutionary history has equipped them to deal with the highly unnatural situation of staying, which gives them a better chance of survival than going.

This example shows how careful we must be in judging animal welfare just by one criterion. The criterion of health and chances of survival would say that the bird's welfare is better in a cage than in the wild. But the criterion of what the animal wants—what it is motivated to do—would lead us to the conclusion that it suffers in a cage, using the definition of being highly motivated to do a prevented behavior. That is why I stressed earlier that we should not rely on just one measure of suffering. An animal may have no obvious signs of ill health or injury, but a healthy animal can still be highly motivated to do something for which, as far as its health is concerned, there is no need but which it still does because natural selection has made it "suffer" if it does not. By keeping animals in conditions that are very different from those in which their ancestors evolved (their danger-avoiding mechanisms, which, as we have seen, include suffering), we may have divorced the danger-avoiding mechanism from the danger the animals were evolved to cope with. We then must discover whether this is, in fact, the case by finding out what a given animal is actually motivated to do or to escape from. We should be most alert to this problem on behalf of the animals we keep in the most unnatural conditions—on farms, in zoos, or even in our own homes. It is to this issue—how we establish the animal's point of view—that I would now like to turn.

We have reached a point in the argument where it is possible to see suffering as a state in which an animal is highly motivated to carry out behavior that it is, for some reason, prevented from doing. The animal may be highly motivated to avoid or escape from, say, a predator or an aggressive member of its own species. In contrast, the animal may be highly motivated to obtain or go toward other things, such as food or a nest box, that it does not have. Either way, we need to find out how motivated an animal is to do something that it is not at the moment doing. If it is only mildly motivated, perhaps we need not worry. If it is strongly motivated over a long period of time, then I suggest we have identified a suffering animal.

Various measures have been used to measure animal motivation or, at least, to infer how motivated the animal must be even when it is not doing the behavior. One of the easiest measures is to suddenly give the animal the opportunity to do the behavior and see how much or for how long the behavior is performed. For example, if battery-caged hens are moved from their wire-floored cages, where they have no opportunity to dust bathe, onto a floor covered with wood shavings or loose litter, they show a great upsurge in the amount of dust bathing they do (Vestergaard, 1980). The "orgy" of dust bathing hens show when the opportunity arises is one piece of evidence that they had, in fact, been highly motivated to dust bathe even when they could not.

A second method makes use of an animal's ability to control its own environment, either naturally or because a human experimenter has given it that opportunity. An accidental use of this method is described by Silverman (1978), who was doing experiments to try to document the effect of cigarette smoke on small rodents such as hamsters and guinea pigs. To do this, he placed the animals in small glass cages and channeled a stream of cigarette smoke toward them down narrow glass tubes. Quite soon, he had to abandon his experiment altogether because the hamsters learned to stop the stream of smoke by blocking the tubes with the only materials available to them—namely, their own feces. The animals were quite clearly demonstrating their "view" of what was being done to them. Smoke—or at least a stream of air—was so unpleasant that they blocked up the only source of air available to them rather than endure it. Words could not have expressed their view any more clearly.

A more systematic investigation of an animal's point of view can be obtained by providing the animal with apparatus that allows it to show how much it values something. For example, many animals express a preference for one environment over another, but the interesting question is not just "What do they choose?" but "Will they pay a cost to get at their choice?" Female rats, for example, will cross an electric grid to reach a male. The strength of the shock needed to stop them shows cyclical changes corresponding to their estrous cycle and is highest during their estrus, or heat. The strength of their motivation to reach the male is fluctuating, as judged by this titration against the pain of electric shock. Cockerels will push open a door to get at a hen, and the strength of their motivation can be weighed against the weight they are prepared to lift (Duncan and Kite, 1987). Similarly, hens will squeeze though a small gap to get at a nest box, and, again, the smallness of the gap can be used as a measure of motivation (e.g., Bubier, 1996).

Of course, there is an element of circularity here. We may be impressed by the fact that a female rat will endure an electric shock to get at a male and conclude that she must be highly motivated because we ourselves dislike electric shocks. How do we know that shocks are as unpleasant to rats as they are to us? To escape from this unsatisfactory circle, we need a measure of motivation in which the animal's choice is pitted not against something we find unpleasant but against all the other things the animal might be doing. To show that the animal is suffering—that is, attempting to give top priority to a particular behavior—we need to show that the animal is more motivated to perform that behavior than anything else. In other words, we need some way of allowing the animal to tell us that what it cannot do matters to it more than anything else. In this way, we use an animal-centered rather than a human-centered way of assessing priorities.

Fortunately, in attempting to do this, we can make use of some terminology and techniques developed by economists for studying human behavior for quite other reasons. Economists make a distinction between necessities (goods that people buy regardless of price) and luxuries (goods they buy only when the price is low). The value of this distinction for our purposes is that luxuries and necessities are defined not by what people say about them but by what they do. People show by their behavior how much value they place on different commodities. If they regard something as a necessity, they show it by buying it even when the price is very high and, assuming that their income is limited, they therefore have to give up other things to be able to afford it. Their demand for necessities is therefore said to be "inelastic" in that it changes little in the face of major changes in its price. Luxuries, by contrast, show "elastic demand"; they disappear from the shopping list when the price gets too high, and the amount of them people buy varies greatly with price. Animals do not, of course, pay money to get what they want in the way that humans do, but they can pay a price in other ways. They can, for example, be trained to press a lever or peck a key for the reward of getting something they want, frequently food, but the reward could also be access to a mate, a nestbox, or the opportunity to dust bathe in soil. The fact that animals learn to do tasks for such rewards in itself tells us something about how the animals regard them. The food, mate, or dust bath must be important enough to the animal that it will work for it. We can learn even more about the value of the reward from the animal's point of view by raising the price and requiring the animal to peck its key or press its lever twice, 10 times, or even 100 times for a pellet. The animal thus has to work harder for what it wants and to trade off time spent working for one thing against time that might otherwise be spent on other activities. Animals may not have a limited amount of money constituting a fixed income, but they have a limited amount of time in each day, and time spent on one activity has to be at the expense of another. If the animal is willing to give up a great deal of time to pressing a lever to gain access to commodity A, even though it has to give up time that would otherwise be allocated to all its other behavior, then it can be said to show inelastic demand for A. The animal shows by its behavior that it is so highly motivated to obtain A that we could begin to infer that it might suffer if kept in conditions where it does not have access to A. In contrast, if the animal shows a liking for A and will press a lever to get it but then refuses to work all that hard when the price is raised, then that the animal may not be particularly highly motivated to gain A and therefore may not suffer unduly if deprived of the reward.

This idea has now been applied to a number of situations that show that animals will work very hard for some things (i.e., they are necessities showing inelastic demand) and will work for other things only if the price is not too high (i.e., they are luxuries showing elastic demand).

For example, Arey (1992) studied sows that were about to give birth and looked at the price they were prepared to pay for food and the price they were prepared to pay for access to straw with which to build nests. When given access to the right materials, even domesticated pigs make quite elaborate nests before giving birth to their piglets. In modern intensive farms, most of them do not have any nest material at all, and so the question can be asked whether they suffer from this lack. Arey built a complex pen in which sows could push one of two panels with their noses. One of the panels opened a door to a side room in which there was plenty of food but no nest material. The other panel opened a door into a second room in which there was no food but plenty of straw. The pigs were thus faced with a choice between food and nest material, although they were free to come out of either of the side rooms and make another choice whenever they wished.

When the sows were two days away from giving birth, their behavior was studied in detail. At this stage, the sows had to press a panel only once to get into the room of their choice. Under these circumstances, they chose to enter the straw room about the same number of times as the food room. When the "price" of one entry into a room was raised to 150 panel pushes, however, the pigs went into the straw room far less often; they did not seem prepared to pay the same price for straw as they did for food. Their motivation for building nests in straw seemed to be considerably less than their motivation for eating food. This fact might, in turn, have implied that straw was not particularly important to them except that their behavior then underwent a very dramatic change as the time of the birth of their piglets drew nearer.

The day before the piglets were due, the sows suddenly seemed to be prepared to work much harder for straw. They would push the panel over and over again to gain access to the straw room. Now, even though both doors required 150 panel pushes to open them, the sows opened the door to the straw room almost as often as they opened the door to the food room. In the 24 hours before they gave birth, then, straw seemed to be a necessity. From the sow's point of view, access to straw for building a nest was as important as access to food.

Another study that used this approach was carried out by Bubier (1996). She looked at the preferences of hens for different commodities such as nest boxes, perches, food, litter, and the opportunity to be near other hens. She then put up the "price" the birds had to pay for these things by making them squeeze through a small gap (hens avoid doing this behavior if possible), or she restricted their "income" by cutting down the amount of time they had available for gaining access to any of them. Both methods showed that the hens regarded some commodities as necessities and would continue to choose them even when the price was high or when the choice involved going without something else. Nest boxes and the opportunity to scratch for food in litter emerged as necessities. The hens thus seemed very motivated to lay their eggs in a proper box and to scratch for food. Other commodities, particularly access to other hens, however, seemed to be luxuries; the hens did not seem to be prepared to pay a high price for them.

There are many other ways of making animals pay a price for what they want—for example, crossing electric grids and enduring blasts of air. Each method gives us an insight into how the animal regards some aspect of its environment by allowing us to see whether the animal will pay the price. This willingness is a preliminary measure of its motivation for different behaviors. The method becomes even more powerful if it is combined with offering the animal a choice and requiring the animal to give up one thing to obtain something else that it wants. In this way, we build up a picture of the priorities that a particular animal has—its motivation to do one thing relative to do others. We learn what matters most to it and what it regards as top priority, as necessities. How does this help us to identify suffering?

I have argued that suffering occurs when an animal is highly motivated to do a behavior that for some reason, such as being confined in a small, bare cage, it cannot do. I have then argued that we can gauge the motivation an animal has to perform one behavior relative to another by identifying which ones show inelastic demand. Does it then follow that animals suffer if deprived of the opportunity to do a behavior that independent evidence has identified as showing inelastic demand? I suggest that this link is a highly plausible one. At the very least, it brings us closer than any other method for identifying the strong negative emotions we associate with suffering.

Of course, it still requires a leap of analogy to say that because we humans consciously suffer when we are unable to carry out top-priority behavior (such as escaping from major sources of pain or being deprived of food when we have had nothing to eat for a long time), other species also consciously experience similar negative states. They could, of course, be "going through the motions" and deciding to do one behavior rather than another all without consciously experiencing anything. But they, like us, have evolved their systems of behavioral priorities and motivation to do some behavior at the expense of all others as an adaptive mechanism to stay alive and reproduce successfully. It is, of course, possible that they could have such mechanisms, show very similar behavior to ourselves, and still lack the element of consciousness, but maintaining this view would require very special pleading. We know that we have conscious experiences of "suffering" when negative emotions such as fear are aroused, and we also know that these negative emotions can act as goads to action, ways of galvanizing us to drop everything and get ourselves out of danger or circumstances that threaten our survival. We show inelastic demand for some things and can, under some circumstances, appear to do anything to obtain what we want. We certainly suffer if deprived of things that are necessities, such as food. It is only a small step to saying that it is very likely that other species do, too.

In other words, we express our point of view by using behavior that is very like that shown by other species. The continuum between displeasure, pain, and suffering, on the one hand, and pleasure on the other appears to be a common and adaptive feature of many organisms (Cabanac, 1992). The advantage of using techniques such as measuring behavioral priorities or the elasticity of demand is that we need make no assumptions about what matters to another species of animal. We let the animals themselves tell us. We thus do not make the mistake of assuming that they are exactly like us but only like us to the extent that an animal deprived of the opportunity to perform high-priority behavior experiences something similar to the unpleasant state we experience when we are prevented for a long time from doing what we very much want to do. What we want and what another animal wants may be quite different (I feel no particular need of a nest box, for example), and to this extent our points of view are quite different. What we share is having a point of view at all.

References

Arey, D. S. Straw and food as reinforcers for prepartal sows. *Appl. Anim. Behav. Sci.* 33: 217–226, 1992.

Broom, D. M., and K. G. Johnson. *Stress and Animal Welfare.* London: Chapman and Hall, 1993.

Bubier, N. The behavioural priorities of laying hens: The effect of cost/no cost multi-choice tests on time budgets. *Behav. Proc.* 37: 225–238, 1996.

Cabanac, M. Pleasure: The common currency. *J. Theor. Biol.* 155: 173–200, 1992.

Damasio, A. R. *Descartes' Error: Emotion, Reason and the Human Brain.* New York: G. P. Putnam's Sons, 1994.

Dawkins, M. S. *Animal Suffering: The Science of Animal Welfare.* London: Chapman and Hall, 1980.

Dawkins, M. S. From an animal's point of view: Motivation, fitness and animal welfare. *Behav. Brain Sci.* 13: 1–61, 1990.

Dawkins, M. S. *Through Our Eyes Only? The Search for Animal Consciousness*. New York: W. H. Freeman, 1993.

Dawkins, R. *The Extended Phenotype*. New York: W. H. Freeman, 1982.

Duncan, I. J. H., and V. G. Kite. Some investigations into motivation in the domestic fowl *Appl. Anim. Behav. Sci.* 18: 387–388, 1987.

Gould, J. L., and C. G. Gould. *The Animal Mind.* Scientific American Library. San Francisco: W. H. Freeman, 1991.

Griffin, D. R. *Animal Minds*. Chicago: University of Chicago Press, 1992.

Krebs, J. R., and N. B. Davies. *An Introduction to Behavioral Ecology,* ed. 3. Oxford: Blackwell Scientific, 1993.

Silverman, A. P. Rodents' defence against cigarette smoke. *Anim. Behav.* 26: 1279–1281, 1978.

Vestergaard, K. The regulation of dustbathing and other patterns in the laying hen: A Lorenzian approach. In *The Laying Hen and Its Environment,* R. Moss, ed. The Hague: Martinus Nijhoff, 1980.

PART IV

EFFECTS ON HUMANS OF RESEARCH INVOLVING ANIMALS

10

ARNOLD ARLUKE
JULIAN GROVES

Pushing the Boundaries

Scientists in the Public Arena

With the growth of the animal rights movement, scientists and technicians involved in animal research have increasingly found themselves to be the target of organized protests and informal social criticism. Rather than enduring emotionally charged and polarized arguments, the biomedical community has sometimes withdrawn from the confrontation altogether or even denied being animal researchers. Not wanting or knowing how to engage their confronters, some animal researchers have simply gone into the closet (Arluke, 1991).

Some members of the research community have, however, pursued a different strategy. In response to the growth and success of the animal rights movement (Jasper and Nelkin, 1992), biomedical researchers have embraced a number of interest groups to counter what they perceive as a growing threat to biomedical research. These groups fall into four categories. First, there are grassroots public relations organizations, such as Americans for Medical Progress, Saving Lives Coalition, and Putting People First. A second category includes patient-originated groups, such as Incurably Ill for Animal Research or Thank You Researchers. Third, there are advocacy groups, such as the National Association for Biomedical Research and its "sister" state organizations, such as the Massachusetts Society for Medical Progress. Fourth, professional associations, such as the American Medical Association (AMA), the National Academy of Science, the National Institutes of Health (NIH), and the American Psychological Association have added their support to animal research.

These proresearch groups have, in their members' own words, sought to provide "public education" and to "get the facts straight" about animal research in a "fight to get the mind of the public." This campaign has taken the form of what Best (1987) calls rhetorical resources—facts, justifications, and policy recommendations—

articulated by scientists, technicians, physicians, biology teachers, and patients in symposia, conferences, seminars, and literature throughout the United States.

Interestingly, the rhetorical resources that the proresearch organizations have employed in these narratives are not just promoting a product or technology, as scientists have done in other controversies (see, for example, Kleinman and Kloppenburg, 1991; Mulkay, 1993). At stake in the animal research controversy is the public image of biomedical researchers themselves, along with the current style of participation and debate over contested issues. Scientists may have their own motivations for speaking out against animal rights activists, but the form and content of the animal research controversy can be understood, in part, by analyzing the implicit objectives of this public relations work.

Public relations specialists manipulate symbols and construct accounts for their clients to influence the public's perception of their client's actions (Jackall, 1988). They know how to make expedient decisions appear altruistic and how to turn weaknesses into strengths and opponents into demons. This work occurs in the context of a diffuse and volatile public, in which everyone is trying to influence their definition of the situation, and reality seems up for grabs.

In this chapter, we argue that, through such public relations efforts, animal researchers are seeking jurisdiction over not only the scientific but also the ethical issues surrounding animal research by giving themselves the moral authority to define acceptable and unacceptable research. This effort comes at a time when increasingly diverse publics, including research scientists, animal technicians, federal authorities, philosophers, animal rights activists, animal welfare advocates, physical anthropologists, animal behaviorists, psychologists, politicians, administrators, and the curious laity, have sought to have a voice in the debate over animal research.

Animal researchers have followed the tactics of other stigmatized groups (Goffman, 1963) to win the moral high ground from animal rights activists and to exclude them from the debate. Their narratives provided researchers with "instructions" on how to regard others and "recipes" for how others should regard them. In the narratives that we studied, the other was differentiated into two categories: the animal rights activists and the general public. On the one hand, research supporters portrayed the animal rights activists as antihuman, deceivers, and terrorists. On the other hand, they viewed the general public as uninformed and too emotional to take part in the debate. Proresearch advocates characterized animal researchers as moderate centrists, heroes, and humans. We conclude this chapter with a discussion about the problems that scientists faced when entering into an overtly political and moral arena.

Research Site and Methods

The data for this analysis come firstly from semistructured interviews with scientists and administrators ($n = 165$) at a variety of research institutions including universities, medical schools, and government research institutes. We conducted the interviews independently between 1988 and 1992, using a snowball sampling technique to generate the final sample. Most of the interviews took place in New England and

North Carolina. Although a few of our informants were hesitant to talk to us, suspecting that we might be infiltrators from the animal rights movement, their desire to tell their side of the story usually outweighed these concerns. Several of our informants, for example, viewed talking to us as part of their mission to educate the public about animal research.

We also attended proresearch workshops and conferences (n = 10), which gave us the opportunity to talk informally with proresearch advocates from across the United States and to observe discussions among them. We learned of these events from our informants, from personal contacts, and through advertisements posted on notice boards in the institutions where we worked. At workshops and conferences, representatives from various organizations gave presentations about the benefits of animal research. In the workshops, trainee animal researchers learned not only the regulations governing the use of animals in research but also about the animal rights movement and strategies for dealing with its proponents. Finally, we collected and analyzed literature issued by the proresearch movement. In some cases, we obtained these from the conferences that we attended. In others, we wrote to the organizations themselves, after learning about them at the conferences and workshops.

The Other

Activists

Proresearch advocates portrayed their most vocal opponents—animal rights activists—as antihuman, dishonest, and violent. While research protagonists were not influencing the content of the debate, their portrayals of activists set the parameters for how concerned parties should express their voices in the animal experimentation controversy; namely, the participants should be rational, honest, and nonviolent. By setting the parameters for debate in this way, proresearch advocates excluded animal rights activists from participating because they would not be orderly and civilized.

Antihuman. Symposia and lectures usually began with speakers—either investigators or administrators—giving a brief description of the animal rights movement as a "lunatic fringe." Activists were painted as antihuman because they insisted that animals were the moral equivalent of humans. For researchers, this moral equivalency was antihuman because it meant that humans would die because animals could not, a concern clearly expressed in an ad that the research community considered running that read: "If animal rights groups have their way, only people will die." Proresearch groups saw this position as so extreme that it disqualified activists from participating in reasonable and rational debate. "Equating the life of a rat or a pig or any other research animal to that of a human is a perversion," an Americans for Medical Progress (AMP) brochure commented. "This insanity must be stopped now."

A common way of alluding to the activists' antihuman stance was to express outrage over their statements or actions. For instance, a printed remark by a member of People for the Ethical Treatment of Animals (PETA) that "We are equally responsible to the rabbit dying in a laboratory as to a child dying in the hospital" raised the ire of at least one proponent of animal research (Paris, 1992a), who wrote: "This attitude of the 'animal rights' movement is a noxious doctrine. It boils down to a corruption of human value. The academic term for it is misanthropy—a disdain for humanity." In other literature, Fred Goodwin (1992), a leading research proponent from the NIH, personally requested that the Pope pronounce the animal rights philosophy as immoral and incompatible with the Judeo-Christian view of humankind.

Almost always, this antihuman position was summarized by evoking at least one quotation by PETA President Ingrid Newkirk. By far, the most common quotation attributed to Newkirk was "a rat is a pig is a dog is a boy," although cited almost as often was her comparison of the killing of chickens with the Nazi Holocaust: "Six million people died in concentration camps, but six billion broiler chickens will die this year in slaughterhouses." Other Newkirk quotes were also used to underscore the activists' "antihuman absurdities" (Hubbell, 1990, p. 72), such as her charge that humans have "grown like a cancer. We're the biggest blight on the face of the earth" and that if her father had a heart attack, "it would give me no solace at all to know his treatment was first tried on a dog."

Speakers highlighted the absurdity of the animal rights position of equating the value of animals' lives with those of humans by showing how the animal rights activists did not adhere to this philosophy in their own lives. At conferences and symposia, scientists found examples of activists being treated for parasites and having difficulties in deciding whether shrimp and insects had rights—suggesting that the activists did not hold all animals to be unequivocally equal to humans. One speaker interpreted Ingrid Newkirk's quote as meaning that she held animals to be *more* important than humans. "I think there's something noteworthy in the fact that the first one she mentioned was a rat," he told an audience of about 100 people. He went on to say that animal rights advocates were misanthropes and that the implications of their views supported the suffering of people:

> One of their slogans, one of their most popular ones that they chant at almost every demonstration, is "Stop the pain and stop the torture." I think that's something that everyone here holds some sympathy for. But the real meaning behind that is to stop all research, no matter what the pain and the torture to our fellow human beings is. . . . The president of Last Chance for Animals says, "A life is a life. If the death of one rat cured all the diseases, it wouldn't make any difference to me. In the scheme of life, we're equal." One rat, all diseases. That's not stopping the pain and stopping the torture. That's continuing the pain and continuing the torture!

At the core of the "antihuman absurdity" was extreme boundary blurring by activists, according to research advocates. To demonstrate the animal rights perspective, one seminar speaker showed a cartoon that had a rodent at the top of a pedestal indicating different life forms. He also showed a slide of someone dressed in a Mickey Mouse costume and said that Mickey was "the quintessential animal that was not an animal but was seen and treated just like us" and that it was this kind of thinking about animals that many people had today.

At the same time that proresearch organizations pointed out the logical and moral problems of the activists' boundary blurring and antihuman sentiment, they drew clear boundaries between animals and humans by affirming animals' scientific purpose. One organization, for example, distributed buttons and pamphlets with a picture of a white rat on them: "Some People See a Rat, We See a Cure for Cancer."

Deceivers. A second way to demonize activists was to portray them as dishonest. One AMA speaker observed, "The worst aspect of the movement, besides lab destruction, is its absolute dishonesty and the outrageous distortions these people engage in." Once seen as dishonest, animal activists could hardly be trusted to engage in genuine debate and should therefore not be permitted to participate. Activists were typically portrayed as spreading what one research supporter called "simple lies" in contrast to the "complex truths" spoken by the research community. In a speech delivered to the Research Defence Society of England, Goodwin (1992) of the NIH noted:

> [The] "stop research" strategy employed by the animal rights movement is the dissemination of falsehoods regarding animal research. And, as de Tocqueville noted "The public will believe a simple lie rather than the complex truth." . . . [A] major problem for the scientific community is that the animal rights groups have been telling simple lies, and we've been responding with complex truths.

Activists spread falsehoods, according to research supporters, by misrepresenting animal research and the researchers themselves. Activists, it was charged, always used out-of-date or sensationalist images that made researchers seem to be torturing animals. Pointing to a slide of an antivivisection ad that featured an apparently uncomfortable monkey struggling in a stereotactic brace, one speaker explained how such pictures were "staged" by activists. Researchers also described how they were portrayed by activists as Nazilike because of their alleged willingness to be cruel. At one conference, a slide was shown of a PETA ad that read, "If you like pulling off wings of a fly you might like a career in biomedical research." Another slide was of an Ingrid Newkirk quote from *Washington Magazine,* where she was reported to have said, "Even painless research is fascism and super-racism."

These misrepresentations extended, according to the research community, to falsifying the actual nature of experiments and their results. "He made me out to be a liar," a speaker at an AMA workshop said of comments by a leading animal rights activist who attacked his research. This speaker also criticized the writing of Brandon Reines, who allegedly claimed that Harvey's heart research did not involve the use of animals in his experiments. He showed how Reines selectively used Harvey's writings and left out explicit references by Harvey to the use of animals. "He falsely represented Harvey's work. This is what you're up against—our opponent is shameless in its presentations."

Activists were also not trustworthy partners for debate because they sometimes cloaked their real feelings about animals to make them more acceptable and moderate in the public's eye. The NIH's Goodwin (1992) observed that because of their "inherent dishonesty," activists could deflect public discussion away from what in his opinion was the key issue: the animal rights philosophy of animals and humans as morally equivalent. Instead, they engaged in "pseudo-medical or pseudo-scientific arguments." Thus, when the physician-founder of the Physicians Committee for Responsible Medicine—the "scientific apologist for PETA"—talks about any scientific aspect of animal research, the credibility of his remarks must be questioned, noted Goodwin: "I object when an activist attempts to hide his or her true position by professing expertise on research, or on the history of medicine, or on the way animal research has or has not contributed to human health."

Such deceit was considered particularly insidious because it was aimed at impressionable audiences. Referring to activists' "propaganda" in schools, one speaker said that it was a "characteristic of fascism to go to the young. It was true of the Nazis, and it's true of the animals rights movement. . . . I am struck by the number of young people who think that science has made the world a worse place. They are disillusioned, one step away from needing treatment" (referring to mental health care). More than just depicting science as harmful or useless, this "propaganda" portrayed scientists as "evil," warned Goodwin (1992):

> In the United States, the bulk of the new investments that PETA is now making is in the schools: Students for Animals Rights. PETA kids. Animal rights music festivals. In Montgomery County, Maryland, where large numbers of the children of NIH scientists are enrolled, the PETA people get into the school under the guise of teaching about pet care. But the real agenda is to portray biomedical scientists as evil.

Proresearch advocates saw students as innocents being led astray by the animal rights activists' simplistic and emotionally appealing lines. The president of AMP (Paris, 1992b), for example, observed that "activists shrewdly manipulate audiences with non-threatening, seemingly logical arguments. Teachers, students and parents are sucked in by their honey-sweet lines and fail to recognize the swarming hive they are stepping into." Paris (1992/1993) also noted:

> Schools and our children have increasingly become part of the political battlefield . . . they [animal rights activists] have launched a planned effort to get to our children before they are old enough to understand the issue, and before they might pursue careers in the biomedical sciences. Children too young to fully understand the moral distinction between humans and animals are returning home from school as vegetarians (which can be physically dangerous for children) and opposed to biomedical research. . . . I encourage you to check your local schools to make sure their curricula has not been corrupted by the "animal rights" activists agenda.

In addition to cloaking their ideological agenda, animal rights activists cloaked their organizational identities, according to proresearch groups. One of the most widely distributed pieces of proresearch literature was an article written by journalist Katie McCabe (1990) for *Washingtonian.* Besides alleging that one of PETA's founders staged photographs of animal cruelty, she presented pictures of PETA leaders in formal dress at a fund-raiser above the caption "The Glitterization of Animal Rights." These pictures were part of a boxed section that claimed that PETA used its funds to pay the legal fees for Animal Liberation Front (ALF) members who had been convicted of university break-ins, rather than its stated goals of education and outreach.

Terrorists. Aside from accusing them of being antihuman and deceitful, spokespeople for animal research further demonized animal rights activists by portraying their tactics as violent and dangerous. According to one AMP brochure, "The more extreme animal rights activists are resorting to terrorism: fire-bombing, vandalism, threats, slander, burglary, physical harassment, and even attempted murder in their effort to hobble the advance of medical science and the saving of human lives." At an AMA workshop, a speaker warned that "science as an enterprise is at risk due to scientific terrorism. Activists' destructive tactics threaten the future of research." In such accounts, animal rights activists fell outside the boundaries of civilized society and could be given no place at the table with reasonable people.

Evidence of violent tactics took several forms in proresearch presentations. Sometimes, simple statistical summaries were given of the number of institutions targeted or broken into. One seminar speaker observed that 54 medical schools have been told they were targeted, and almost 4,000 incidents have occurred at medical schools and researchers' homes. At other times, visual evidence of break-ins was shown as part of a slide show or a video. Slides shown by one speaker included a newspaper article about an ALF arson attack at a University of California, Davis, lab that caused $4 million in damage; a picture of wall at a Northwestern University lab—"a graffiti attack"—saying "animal researchers are corrupt and perverse," "animal researchers are Nazis," and "Larson you are a dead man"; a Louisiana State University lab wall sprayed with "Cat killer, go to hell" because the researcher, who had regularly received numerous threats, was "under attack"; and a postcard the speaker received that said he was a "psychopathic swine," with a photo of him with a skull drawn over his face and a death threat written under it.

One proresearch video included still photographs of swastikas painted on lab walls by activists. In an animal training workshop at a government research institute in North Carolina, speakers showed their trainees a video made by a television company in San Francisco, in which the reporters described the animal rights movement as "a group gone violent." They showed clips of burning buildings—fur-retailing stores and animal research laboratories in London—that had been attacked by animal rights activists. The footage showed journalists interviewing an ALF member in what they described as "a seedy hotel." The interviewee's face was masked with a black stocking, and his voice was disguised to preserve his anonymity. This video, according to the speaker at the workshop, "adequately reflects the kind of environment in which we find ourselves using animals in research today."

On other occasions, proresearch advocates simply quoted activists as proof of their terrorist intent. One AMP brochure, for instance, cited Tim Daley of the ALF, who purportedly said, "In a war, you have to take up arms and people will get killed. . . . I can support that kind of action by petrol bombing, bombs under cars, and probably, at a later stage, shooting [researchers] on their doorsteps."

Other forms of evidence included anecdotes by well-known scientists of attacks against them. One celebrity victim, for instance, made the focus of his speech the four times he had received bomb or death threats. Another researcher talked about how the bomb threats he got "had a devastating impact on my life," making him live in fear for his life and the lives of his family. In some of these anecdotes, threats were made against scientists not only because of the nature of their research but also because they had spoken out against activists. Speakers acknowledged that many of their peers were uncomfortable talking at public functions because they felt that they were not trained to speak in this capacity and because they were afraid that such speaking would make them targets. At one seminar, a researcher said:

> I'm here as the designated attackee. I was attacked. I was attacked, not just because of my research, but because I had been speaking out on behalf of animal research, and noting the dangers posed by the animal rights movement. . . . ALF was in my office one month later. They looked for proof that my speaking was supported by the government, but they couldn't find any link. But I am now working for Fred Goodwin. I get the usual death threats and hate mail.

In their effort to label activists as dangerous, proresearch groups have even pushed to legally classify laboratory break-ins as terrorist acts. A politically conservative consulting firm produced a report (Hardy, 1990), supported and distributed by the AMP, that catalogued the "terrorist" activities of animal rights groups and attacked the reticence of federal authorities to enact legislation to deal with such terrorism. This report, like the firm's other publications, was written to be cited and relied upon for court decisions.

The General Public

Proresearch advocates distinguished "the general public" as a second critical voice in the controversy. Research supporters variously referred to this category as the "mainstream American housewife," "the average person," "high school students," "the media," "the lay public," "Americans," or "society." Although proresearch narratives did not demonize the general public or flatly rule out their participation, their voice was still problematized. Unless they could be taught to think and feel like scientists—tempering their emotions and becoming more informed and rational—they were also disqualified from participation.

The proresearch contingent questioned the capacity of the public to make serious contributions to the debate by portraying them as ignorant about science and animals,[1] which rendered them ill equipped to advise on any policy having to do with the use of animals in laboratories. At an AMA seminar, one speaker lamented the antiscience attitude of certain segments of the population. He noted that "ignorance is the problem—no knowledge of the scientific method." The antiscience minority, in his opinion, were "contemporary Luddites." He cited a 1957 study by Margaret Mead of 35,000 high school students' stereotypes of scientists; one element was that students saw them as cutting apart animals and injecting weird serums. Similarly, another speaker bemoaned the "worrisome trend" toward a growing antiscience movement and science illiteracy. At another conference, speakers expressed concern about survey data indicating that almost all American schoolchildren were scientifically illiterate. "They don't know which comes first, thunder or lightning," observed one investigator. Another speaker talked about his dismay when he discovered that none of the high school students with whom he had spoken knew what an iron lung was.

Speakers identified several aspects of the public's ignorance that were thought to disqualify them as partners in debate. Being uninformed about science and animals, they would be too confused, amateurish, or irrational to seriously address questions of animal experimentation and to carefully weigh the costs and benefits of research.[2] One speaker, for instance, noted that because only 4 percent of the American population now lived on farms, many people held a "romanticized notion of animals based on our experiences with pets." Because the lay public had such limited experience with animals, their understanding of them would be distorted, which would lead to inappropriate or wasteful suggestions, according to researchers, especially when this lack of experience was combined with the public's emotionality over animals. For example, at a workshop on primate well-being, researchers lamented lay suggestions to increase cage size because they were based on anthropomorphism rather than scientific understanding of the psychological needs of primates. Another speaker was troubled that many people lumped together cosmetics testing and biomedical research.

By far the most commonly noted sign of public ignorance, however, was not knowing the role that animal research played in achieving medical advancements. Because the public was unaware of the contributions of animal research, they were not qualified to sit on regulatory committees. The chair of one such committee told us: "The average person doesn't have all the information or does not spend the time to think about the implications of animal-based research, the implications of the loss of animal-based research, and therefore society's views are perhaps somewhat less relevant than our own in that respect." Some research supporters attributed this ignorance to the fact that Americans did not pay enough attention to science because they were preoccupied with "pseudoscience" or other frivolous activities. One speaker, for example, cited the popularity of astrology columns in the newspaper as evidence. Another complained that Americans were more interested in sex and entertainment on commercial television—"the idiot box"—than the educational documentaries that the public television station had to offer: "One of the problems is: who watches Channel 4? Very precious few people. You get some marvelous programs on public television about biomedical research, discussions about ethics and all of this stuff. Who watches it? Nobody. It's not sexy. It's not interesting. It's not entertaining. It's not fun."

The Self

Scientists

Proresearch narratives excluded opponents by problematizing them. Either they were demonized, or stringent conditions were set for their participation. Another way in which they excluded critics was by establishing the legitimacy of scientists' voice in the debate. In this set of narratives, proresearch spokespeople portrayed the research community as moderate centrists, heroes, and humans.

Centrists. To expel activists from participation in the controversy and, at least in theory, to include a certain segment of the public in the potential debate, the research community sought to define itself as moderate centrists rather than as the mirror extreme of activists. Without such positioning, their voice would also lack credibility in the debate. Presumably, in this middle ground, the research community could enter into debate with the concerned public.

By finding quotes from animal rights activists to support the "total abolition" of all animal experiments, the proresearch community (e.g., Horton, 1989, p. 742) described how the animal rights movement represented an extreme or radical position. As one speaker remarked, reiterating Ingrid Newkirk's words, "The animal rights movement is not interested in bigger cages but abolishing all cages." Real dialogue, research proponents argued, could not take place with those who sought such total abolition, and only with those critics who were more moderate.

To posit themselves as a party willing to debate in a reasonable manner, research advocates sought to define the middle ground. One speaker reminded his audience that researchers were squarely in the middle ground because the real spectrum was animal rights at one extreme and willful mistreatment of animals at the opposite extreme. The problem, he said, was that because willful mistreatment did not exist, then research itself became falsely painted as the extreme. According to another speaker, the middle ground had to do with how animals were used, not whether they were used. In this regard, he noted, "The research community has moved very definitely into the middle ground" because it neither advocates "an absolute commitment to the animal's comfort and procedural safeguards" nor a blatant disregard of the well-being or suffering of animals. This statement loosely positioned the biomedical research community with traditional animal welfare organizations. "As scientists, we can consider ourselves as protectionists. We care for animals," a speaker from the NIH noted.

To be positioned with animal protection groups was, however, a delicate matter. The literature distributed by the research community attempted to sort the good from the bad animal protection organizations. They frequently described how the old traditional animal welfare organizations, who truly represented public opinion, were being taken over and radicalized by the leaders of the animal rights movement, who were consequently acquiring control of larger sums of power and money. Referring to the American Society for the Prevention of Cruelty to Animals (ASPCA), for example, Goodwin (1992) of the NIH observed that it had shifted toward the animal rights end of the spectrum, "but, thankfully, many local community charities in the United States remain true to the principles of animal welfare, and have not been corrupted by animal rights groups." This advocate went on to say: "In the past, the ineffectual response of the scientific community and others to the animal rights challenge contributed to a gradual radicalization and takeover of many traditional animal welfare organizations. It is no longer easy to determine whether some humane societies are in the welfare position or the rights position."

To buttress their case for centrism, proresearch groups frequently cited attitude surveys of the American public and the medical profession that suggested the research community was not a minority voice, let alone an extreme one. For example, AMP's literature, along with other proresearch literature and presentations, frequently made reference to a survey claiming that 77 percent of those polled agreed that the use of animals in research was necessary for medical progress. The AMA (1992) claimed that "polls show that the American public overwhelmingly rejects the activists' claim that there is no difference between animals and humans. So, where does that leave the animal rights movement?"

Heroes. The proresearch literature and symposia attempted to reverse the negative image of animal researchers and their work by giving triumphant accounts of the contribution of biomedical research to medical progress. More than merely extolling the benefits of research, these accounts characterized investigators as unsung heroes whose future achievements were jeopardized by the growing voice of animal rights activists. Proresearch speakers would, for instance, often begin their presentations by asserting that medical research has added 26 years to our life span this century. The first pages of a prepared speech for the public (AMA, 1992) noted:

> You've heard people wish for "the good old days." We all know what they mean by it, of course, but the fact is, the good old days weren't all that great for most people. For example, the average life expectancy of a person born in ancient Greece or Rome was about 20 years. A couple of thousand years later, things still hadn't improved much. In 1850, the average life expectancy was about 40 years—at least in places like Massachusetts, which was probably one of the more healthful areas of the world. In another 50 years, by 1900, the average lifespan of a person born in this country had crept up to 47 years. What about now? The average American born in 1990 will live nearly 76 years—and enjoy a healthier life than his or her ancestors. It's easy to forget how much our medical well-being has improved.

After assertions were made about increased life span, presentations litanized "one-time scourges" purportedly defeated through animal research, which usually included anthrax, cholera, smallpox, measles, tuberculosis, leprosy, tetanus, diphtheria, rabies, whooping cough, and hepatitis B, as well as some diseases that were "brought under control so long ago that their names are now quaint and unfamiliar," like pellagra, rickets, or beriberi. The problem, these spokespeople agreed, was that these medical achievements had been taken for granted. A surgeon at a symposium described his experience of practicing in third-world countries as "a glimpse into the past" without biomedical research. "Flocks of children," he told the audience, "regularly die of smallpox and measles." Another surgeon even admonished his own colleagues for forgetting medicine's contributions:

> We do not appreciate how far we have advanced in the last century, and how important animal models have been in achieving these advances. Who among us has ever seen anyone die of whooping cough, measles or polio? We take for granted these vaccines that were developed with animals and enabled these life-threatening infectious diseases to be brought under control. Most current practicing physicians don't know how lethal common bacterial infections were prior to the advent of penicillin and of many generations of antibiotics that have since been developed and tested in animals before administered to humans. We've forgotten that people in our grand-

> parents' generation frequently died from complications of appendicitis and communicably acquired pneumonias.

Of all the illnesses cited, polio received the most rhetorical attention. One brochure put out by a research-based pharmaceutical company in North Carolina included a black-and-white photograph of a hospital ward with more than 50 iron lungs neatly aligned. Such a photograph, supplied by the March of Dimes Birth Defects Foundation, might have been used as a symbol of the conquest of medical technology in preserving human lives in its day. In the context of this literature, however, it symbolized the miserable conditions of human existence prior to the development of the polio vaccine through animal research. "In the 1940s and '50s," the caption read, "polio wards such as this one were commonplace in the United States, where 33,000 polio cases were reported in 1950 alone. Animal research was vital to the development of vaccines that brought the disease under control in the 1960s." The conquest of polio served as such an important symbol of researchers' triumphs that the Coalition for Animals and Animal Research (CFAAR) used a life-sized iron lung to counter an animal rights demonstration. One of this organization's newsletters (Russell, 1990) described how participants celebrated the anniversary of the introduction of the Salk polio vaccine by rolling the lung into a shopping plaza where animal rights activists were staging a protest against research:

> Rolling over the concrete toward the Plaza, gaining momentum, the iron lung made quite a dreadful noise and attracted considerable attention . . . a young woman asked a CFAAR demonstrator what gave people the right to use animals. "What makes us special?" she asked, in a voice that clearly indicated she thought the answer was "nothing." The CFAAR member turned to the iron lung and declared, "This is what makes us special. Only humans could invent a device like this, that saved hundreds of thousands of lives, and only humans could have developed a vaccine to conquer polio, to make the device unnecessary."

Although these success stories made researchers look like heroes who worked "miracles," this image was more easily sustained by relying largely on past successes, such as the smallpox or polio vaccine. Because there were fewer recent "magical bullets" to cite, advocates relied far less often on more recent biomedical research, described by some as "half-way technologies" (Thomas, 1977), given their expense and their failure to provide true prevention or cure.[3] Proresearch advocates dealt with the problem of not achieving recent and continuous magic bullets by suggesting that scientists were "on the verge" of achieving breakthroughs with many major diseases. For example, AMP's newsletter, *Breakthrough,* was devoted to publicizing the advances made by researchers for problems such as ulcers, migraines, cystic fibrosis, cancer, arthritis, and Alzheimer's disease. Reports typically focused on admittedly small, although necessary, gains in basic research that presumably would lead, one day, to actual cures. Another way that they dealt with this problem was to intertwine discoveries of vaccines and cures with diagnostic advancements and even current research, thereby blurring the line between magic bullets, half-way technologies, and the promise of experiments.

After investigators made their presentations at symposia, patients themselves were often relied on to extol the benefits of research. Organizers encouraged testimony from those who were the supposed beneficiaries of animal research. One man, representing an organization called Incurably Ill for Animal Research (IIFAR), explained at a symposium how he had been severely burned in a plane crash and treated with surgical techniques and drugs that he attributed to animal research. During his presentation, he made a point of thanking the research community:

> Yes, the benefits of research are very real. And to some of us they are evident every day. And we're grateful for the research that's been done in the past that's helped us. We're grateful for the researchers and the rest of the research team that has made all of research possible. And I know that we have some members of the research community here with us tonight. And I'd like to take this opportunity to say thank you for myself personally, and the rest of our members. 'Cause I think that's something that's not said up 'til now.

Such testimonials served as reminders to researchers, and as notification to the general public, that animal research was appreciated, at least by some people. Indeed, a proresearch organization was recently formed with the name "Thank You, Researchers." More than merely thanking researchers, their speeches told, almost in a religious sense, of how they had been "blessed" by the researchers who had worked "miracles" by making it possible for people like themselves to be "reborn." Who could question the loss of a few animals, they argued, given the incredible benefits of their research. To wit, a woman who had received both a liver transplant and a hip replacement spoke at one seminar about her experience and gratitude to researchers:

> I have been blessed by every member of this room. I am extremely grateful to all of you. We must keep in mind the main thing—biomedical research saves human life. I had primary biliary disease. I was told I would die in two years, but I'm still here. You are looking at a genuine miracle, but miracles don't happen. [It]'s what science and medicine are all about—to discover wondrous miracles. I am grateful, thankful, and indebted to all of you. Thank you, thank you, thank you. And thank you on behalf of millions of people who depend on your skill. I am here, alive and well, because I have benefitted from scientific investigation involving the use of animals. I watched my son graduate from law school, I celebrated my 35th anniversary, but 10 years ago I lay comatose and dying. My family was told to plan my funeral, write my obituary. [She then tells a story about how she acquired someone's liver.] It saved my life. I am alive and well. How can I describe what it feels like to be reborn? How can I explain to you how grateful I am to you for each new day? How truly terrified I am that your vital research will be deterred by misguided animal rights activists and that other people will be deprived of their rebirth. Those who impede medical research do a disservice to themselves. We cannot afford to put one medical miracle at risk.

These testimonials, as well as presentations by investigators, frequently ended on notes of both optimism and pessimism. On the one hand, presenters were optimistic about the possibility of developing treatments for Alzheimer's disease and AIDS through animal research in the future. On the other hand, they added an apocalyptic warning that because of a "misguided," "overzealous," and "fanatic minority," we risked "throwing it all away" and "depriving our children" of the benefits of such research. Thus, the heroic narratives posited the animal rights movement as the primary obstacle to new medical miracles. The possibility that such triumphs might be curtailed by activists was emphasized in the very title of one all-day workshop, "Medical Progress—A Miracle at Risk," sponsored by the AMA, the Massachusetts Medical Society, and the Massachusetts Society for Medical Research. These miracles would be in "danger" and "at stake" (AMA, 1992) if the biomedical research community's voice did not prevail in the animal experimentation controversy.

These rhetorical devices again precluded the participation of the animal rights voice from the controversy because they shifted attention away from philosophical arguments about the value of animals' lives and accusations of cruelty on the part of researchers to the positive outcomes of the research itself. Such narratives focused on the ends rather than the means of animal research. The ethical choice was not whether to use animals in research but whether to deprive those who are sick and vulnerable of its fruits. "Is it ethical to let a child die of a disease when you know that animal research can save her?" an IIFAR speaker asked rhetorically at a proresearch symposium.

Nevertheless, through another set of claims, proresearch advocates also sought to address issues relating to the methods of research, by showing the public that these methods were noble, too. They addressed these issues by presenting themselves as "humans."

Humans. To be effective with the public, proresearch groups encouraged scientists and technicians to present themselves in a personal and egalitarian manner, as opposed to detached, elitist scientists. They advised animal researchers to show that they did have feelings for animals and were working in the interests of patients who would benefit from their knowledge. To this end, proresearch journalist Katie McCabe (1987) gave a speech at a convention for the California Biomedical Research Association (CBRA)—excerpts of which were distributed by the similar organization in North Carolina—in which she advised researchers:

> If you can't humanize your story, reporters won't write it and editors won't buy it. . . . Know that this is not a clean, fair fight. Those who go in expecting a reasonable debate will lose out to shrill, emotional arguments and ad hominem attacks. Yes, you must address the animal activists' arguments about alternatives, duplication, etc., but you will only win by humanizing your viewpoint—by being a human being first concerned about other human beings. Be a human being first, a scientist second.

After describing how a physician introduced her, by name, to his patients and how an animal researcher convinced her that his research promised cures of Parkinson's disease, she told researchers:

> Suddenly, the issue of animal research became very personal; it had a face and a name. . . . I had what every reporter wants: a people study. . . . The human element was critical; the scientist became the man next door, a man who admitted to conflicted feelings about animal research, but who had chosen to continue doing it because of the human benefits. . . . Like it or not, this is an emotional issue: it is a public relations battle for the hearts and minds of the public, and, as uncomfortable as this makes you as scientists, I believe people's hearts must be won before they will listen to the substance of the arguments.

Scientists were encouraged to show feelings for lab animals or pets,[4] thereby not completely ruling out the place of emotions in the controversy. But this emotionality was different from that of activists or the laity because it was grounded in experience with animals and science and tempered by rationality. A speaker at one seminar acknowledged that researchers might be hesitant to do this: "I know it sounds heterodox to say, but we should express our concern. Some of you will think this is like getting in bed with the enemy, but mostly people want to know that you care about animals—so say you have pets and care about animals."

Indeed, scientists talked about their affection for their own pets, as well as for lab animals. At the symposia and public hearings, researchers declared laboratory animals "partners in research" and even publicly thanked them. A pediatrician told his audience: "I close with a word of thanks to those dogs who provided what was necessary for my 89-year-old mother to have both her hip joints replaced." Testimonies by surgeons and veterinarians emphasized the direct benefit of animal research to animals themselves. At one symposium, a veterinarian explained how animal research had protected animals from rabies, feline leukemia, canine parvo virus, and distemper. At the AMA workshop, a Nobel laureate spoke about his research on kidney transplantation, saying "I love animals. I have a dog and a cat at home, my grandchildren also do." He went on to talk about the dogs he used for his transplantation research and showed a slide of one of these dogs seated behind its birthday cake. "They became pets for all of us. Many were taken home for family pets." Those who attended the workshop were also given, in their resource kits, a prepared talk with accompanying slides. Slide 36 (AMA, 1992) reiterated: "Scientists aren't white-coated Frankensteins. They are literally the people next door. They have families and pets. They have sons who love their dogs, and 12-year-old daughters who are crazy about horses. Their compassion is the first line of defense against cruelty."

To further humanize their image, research supporters emphasized how medical researchers were motivated by the altruistic goals of helping patients. No doubt, the proresearch community promoted this image to counter animal rights accusations that they undertook unnecessary research to promote their own careers, but it also served to soften the public's view of scientists as cold, insensitive intellectuals. At the core of this image was the wish to make the proresearch rhetoric more emotionally appealing by portraying scientists as rescuing victims, usually shown as healthy and appealing young people and animals, from the ravages of disease. The CFAAR members in California thus wrote in a letter to the editors of *Science* (Denver, Nicoll, and Russell, 1988):

> We have learned that when we are dealing with the activists and with the public, the most effective approach is not only to use rational dialogue, but also to make the topic emotionally appealing. We therefore show photographs of young, healthy children; healthy, elderly persons; healthy cats and dogs; and include a brief statement about how animal research benefitted them.

They acknowledged, however, that most researchers would find this difficult to do. As the NIH's Goodwin (1992) commented:

> It is difficult to get academic researchers to over-simplify and to infuse their arguments with emotion. Goes against our grain. We tend to be logical, reflective, and careful, and not to overstate anything. But in a political battle with a radical group, a purely intellectual approach is ineffective and even counter-productive. . . . We have to develop brief, emotionally engaging examples of the medical benefits of animal research, in terms understandable to the public. Medical research is an esoteric, complicated field, and most scientists are not particularly good at public education; that is not why they went into science.

At an AMA workshop, two public relations experts told an audience of scientists, physicians, and technicians to "talk about your research to people with the same passion you talk about your patient care." Following this advice, physicians invited to speak on behalf of animal research began their presentations at symposia by saying that they were going to share "personal interfaces" or "share some stories." *The Compassionate Quest,* a brochure distributed by the North Carolina Association for Biomedical Research, quoted a research physician as saying: "When one of my patients dies, all I can do sometimes is come back to my office and cry. [It]'s where the issues of animal studies gets very clear for me. I don't like to see an animal die. But I hate to see a child die."

Slides and color pictures in brochures often depicted physicians bent over or helping children. Short narratives, both in their literature and at symposia, described children on the verge of death brought back to life by physicians who used techniques developed from animal research. In one brochure, a child named Charlotte was born with "spaghetti-thin arteries in her lung." It was "a race against time" until doctors perfected the heart and lung transplant and the drugs necessary for her survival. Eventually, Charlotte's situation was ameliorated when, after performing the technique on lambs, a doctor performed a balloon angioplasty on her pulmonary arteries. At a symposium, the speaker representing IIFAR played a video of her mother explaining, "My daughter was in that doctor's hands, and he brought her back. Surely it's worth a few laboratory animals."

Proresearch advertisements and pamphlets characterized children as the potential victims if animal experimentation were to be stopped.[5] In one AMP ad in the *New York Times* and the *Washington Post*, a photograph of a smiling, attractive, healthy-appearing young girl was featured above the caption "She May Owe Her Life to a Rat, a Monkey, and a Lot of Dedicated Scientists." The text noted: "There's good news for the 30,000 young victims of cystic fibrosis. They can now hope to live a long time. Dreaded genetic disease . . . may soon go the way of diphtheria, smallpox and polio." A poster put out by the Association for Research in Vision and Ophthalmology showed an apparently healthy infant and its head being cradled in human hands. "Her Eye Disease Comes from Birth. Her Hope, from Animal Research," the bold print read.

While children were often portrayed as victims, proresearch groups portrayed scientists as their rescuers. A Foundation for Biomedical Research (FBR) poster, for example, read, "If We Stop Animal Research, Who'll Stop the Real Killers?" Beneath the large print, three large microscopic pictures were shown of cancer, heart disease, and AIDS.

In other scenarios, proresearch presentations and literature portrayed the scientists themselves as victims—victims of animal rights activists. Scientists were in danger, it was argued, of becoming the prisoners of extremists. An AMP booklet was titled "Assuring the Freedom of Medical Science in Its Continuing Quest for New Cures." Reprints of a *Reader's Digest* article (White, 1988) that AMP mailed to interested parties referred to scientists as "shackled" and "straitjacketed." One AMP poster showed a man's hands lashed together at the wrists by rope with the bold-print caption reading: "What Are They Doing to Our Research Scientists?"

In sum, researchers not only promoted the benefits of animal research but also their humanity and benevolence in bringing about those benefits. By emphasizing their affection for animals and children, the researchers attempted to show the public that they had not lost touch with their emotions and were thus sensitive, competent participants in the debate. The "real killers," as the poster suggested, were the diseases that the researchers fought. At the same time, proresearch accounts implied that animal rights activists were not aware of the hardship of human illness, and therefore their preoccupation with animal suffering was misguided. Lost in heady philosophical ideas, they did not inhabit the real world and therefore did not appreciate the difficult choices that medical scientists had to make. Referring to animal rights philosophers, a veterinarian in a medical school told us:

> It must be nice to have chosen a profession that is totally dealing with words and ideas. I think all you have to do is see somebody in the burn center or watch your wife go through an appendectomy as an adult and have the surgeon tell her a fair number of adults die from appendicitis to feel a commitment to humanity and a little less strong commitment to animals.

Discussion

The animal experimentation controversy illustrates how medical scientists have sought to capture the moral high ground in ethical decisions by promoting not only a technology but also a moral identity that is superior to their opponents. In presenting themselves as moderate centrists, heroes, and humans and the animal rights activists as antihuman, dishonest, and terrorists, the proresearch movement has attempted to exclude others from having a voice in the controversy over the use of animals in biomedical research.

How successful the proresearch movement will be in winning jurisdiction over decisions surrounding animal experimentation remains to be seen. If nothing else results, however, the narratives of proresearch groups may be instrumental in creating a subculture among researchers: the creation of an identity that is lacking. In an age in which science is frequently under attack, scientific progress is incremental and increasingly specialized, and vivisection is becoming morally tainted, these researchers learned how to translate their uneasiness into confidence. One young graduate student, for example, described her local CFAAR chapter as a "support group." She went on to tell us: "Before we formed, there was really no group on campus that was proresearch. We assume that most of the scientific community is proresearch, but they don't have any formal group to say that they are."

But proresearch organizations present researchers with new dilemmas. Scientific authority rests on clear boundaries that distinguish scientists from the laity (Gieryn, 1983). Such boundaries have traditionally sought to demarcate scientists as neutral rather than judgmental, rational rather than emotional, and autonomous rather than malleable to popular opinion. If what distinguishes animal researchers from animal rights activists and the public is autonomy and an unemotional persona, then the formation of interest groups and humanistic and emotional presentations can blur these distinctions. New boundaries may have to be created. Indeed, animal rights activists have exploited this blurring as an opportunity to label the researchers as the emotional and insidious party in the controversy. At one proresearch rally, animal rights activists staged a counterprotest. One activist insisted, referring to the proresearch literature, that the scientists were the harbingers of sentimentality and misinformation in the controversy: "It always amazes me that our opponents accuse us of being overly sentimental, of being misinformed . . . yet the three pictures that they are always working with, and that you see here today, are, in my humble opinion, classic examples of mushy sentimentality and distorted propaganda!"

As with any subculture, not all stigmatized people share in it (Warren, 1974). Some, as we noted, withdrew from the controversy. A few of our informants described anguish about having little experience in appearing on television and concern that the academic community was being dragged onto "the street corner" along with the animal rights activists. Then again, some enjoyed the challenge of debating the activists and expressed dissatisfaction with colleagues who failed to do so. One of our informants, a researcher-physician, was the son of a preacher. He told us about how he enjoyed making public presentations and viewed challenging animal rights activists, along with antiabortion activists, as "a sport":

> I used to make a great sport of walking into the hospital through the front door so that I could confront one of these individuals [antiabortion activists] and look them in the eye and say: "Tell me, how many adopted children do you have?" I have never met one who had an adopted child. And I feel much the same way about these people [animal rights activists] that I characterize as "humaniacs."

As sociologist Erving Goffman (1963, p. 27) observed with respect to those who make careers of their stigma: "Instead of leaning on their crutch, they get to play golf with it."

ACKNOWLEDGMENTS We appreciate Andrew Rowan's comments on an earlier draft of this paper.

Notes

1. Certainly, animal rights activists were also characterized by research advocates as ignorant about animals. For example, a popular article frequently cited in the proresearch literature (Conniff, 1990) discussed how the animal rights movement had elevated abysmal ignorance about animals to a philosophy. But the problem of ignorance about animals and science, according to the research community, extended far beyond activists to the country at large.

2. Public ignorance was also worrisome to researchers because it was seen as a window of opportunity for activists to make gains. The poor performance of American students on science exams made them a likely "target" for activists. "A young public that knows or understands virtually nothing about science offers a tremendous target of vulnerability for anyone who wants to distort or discredit science," observed one advocate (Goodwin, 1992, p. 3).

3. In the late nineteenth century, the medical community also sought to win public support by exaggerating the efficacy of treatments that came out of animal research. According to Turner (1980), claims were made that treatments had cut the mortality rate of diphtheria by as much as 40 percent when, in actuality, figures were closer to 10 percent.

4. Many other emotions for lab animals were expressed in private. For example, researchers claimed that, because it would upset them, they would not work with certain animals or do certain procedures. Despite these feelings, most researchers did not hold it against others who could do these experiments.

5. Both the animal rights movement and the proresearch movement derive meaning from their respective symbols in similar ways. Both movements key into symbols of innocent victims. In the former case, animals are the victims of scientists, while in the latter case scientists, if not the entire society, are the victims of activists. Both movements also cast their relationship to victims in heroic terms. Activists talk of rescuing lab animals from their plight; researchers speak of saving children and future generations.

References

American Medical Association. *A Miracle at Risk.* Chicago: American Medical Association, 1992.

Arluke, A. Going into the closet with science. *J. Contemp. Ethnog.* 20: 306–330, 1991.

Best, J. Rhetoric in claims-making: Constructing the missing children problem. *Soc. Prob.* 34: 101–121, 1987.

Conniff, R. Fuzzy-wuzzy thinking about animal rights. *Audubon* 92(6): 120–132, 1990.

Denver, R., C. Nicoll, and S. Russell. Direct action for animal research. *Science* 241: 11, 1988.

Gieryn, T. Boundary-work and the demarcation of science from non-science; strains and interests in the professional ideologies of scientists. *Am. Sociol. Rev.* 48: 781–795, 1983.

Goffman, E. *Stigma: Notes on the Management of Spoiled Identity.* Englewood Cliffs, N.J.: Prentice Hall, 1963.

Goodwin, F. Animal research, animal rights and public health. *Conquest* August (181): 1–10, 1992.

Hardy, D. *America's New Extremists: What You Need to Know About the Animal Right's Movement.* Washington, D.C.: Washington Legal Foundation, 1990.

Horton, L. The enduring animal issue. *J. Natl. Cancer Inst.* 81: 736–743, 1989.
Hubbell, J. The "animal rights" war on medicine. *Reader's Digest* 136 (June): 70–76, 1990.
Jackall, R. *Moral Mazes: The World of Corporate Managers*. New York: Oxford University Press, 1988.
Jasper, J., and D. Nelkin. *The Animal Rights Crusade.* New York: Free Press, 1992.
Kleinman, D., and J. Kloppenburg Jr. Aiming for the discursive high ground: Monsanto and the biotechnology controversy. *Sociol. Forum.* 6: 427–447, 1991.
McCabe, K. The Growing Power of the Animal Rights Movement. Speech given at the meeting of the Board of Governors of the California Biomedical Research Association, 1987.
McCabe, K. Beyond animal cruelty. *Washingtonian* 25(5): 72–77, 185–195, 1990.
Mulkay, M. Rhetorics of hope and fear in the great embryo debate. *Soc. Stud. Sci.* 23: 721–742, 1993.
Paris, S. Lives aren't the same. Letter to the editor. *Belleville News-Democrat,* Belleville, Illinois, 23 July, 1992a.
Paris, S. A rat is not a pig is not a boy! Letter to the editor. *Wall Street J.* 7 October, pp. A15, A17, 1992b.
Paris, S. Letter from the president. *Progress* 1(2): 1, 1992/1993.
Russell, S. CFAAR "celebrates" world lab animal liberation week—with an iron lung! *CFAAR Newslett.* 3(1): 1–3, 1990.
Thomas, L. On the science and technology of medicine. In *Doing Better and Feeling Worse,* J. Knowles, ed. New York: W. W. Norton, 1977.
Turner, J. *Reckoning with the Beast: Animals, Pain, and Humanity in the Victorian Mind.* Baltimore: Johns Hopkins University Press, 1980.
Warren, C. *Identity and Community in the Gay World.* New York: John Wiley, 1974.
White, R. The facts about animal research. *Reader's Digest* 132 (March): 127–132, 1988.

11

HAROLD A. HERZOG

Understanding Animal Activism

Over the past 25 years, the animal rights movement has emerged as a major social and political force in the United States, Great Britain, and some parts of continental Europe. Animal activists argue that the exploitation of nonhuman animals at human hands, including their use as subjects in behavioral and biomedical research, is immoral. Behavioral experimentation on animals has come under particularly heavy criticism from animal protectionists (e.g., Rollin, 1989; Singer, 1975; Shapiro, 1997). The targeting of behavioral researchers by animal rights proponents is somewhat ironic in that some individuals who have chosen to devote their professional lives to discovering why other creatures do the things they do sometimes share with their critics a love for animals and a respect for the natural world.

Many animal behaviorists, particularly those with an ecological-ethological orientation, have been sensitive to changing attitudes toward the use of research animals. The Animal Behavior Society in the United States and the Association for the Study of Animal Behaviour in Great Britain have been relatively advanced among scientific organizations in establishing guidelines for the use of experimental animals. After surveying the editorial policies of biomedical and behavioral research journals, Orlans (1993) concluded that the journal *Animal Behaviour* was among the most progressive in its policies concerning animal welfare. This is not to say that ethologists and behavioral ecologists are in agreement as to the moral status of animals. On the contrary, the animal behavior research community does not appear to have a consensus on this issue. (See, for example, the debate between Bekoff [1993] and Emlen [1993] over the ethics of experimentally induced infanticide.)

Researchers sometimes seem baffled about how to effectively discuss issues related to the ethical treatment of animals. Some continue to wonder what the fuss is all about. Many scientists, including animal behaviorists, are not aware of the sociological, psychological, and philosophical underpinnings of the animal protection movement. Animal activists are frequently dismissed as anti-intellectual misanthropes who prefer kittens to sick children and who conveniently ignore the contributions of applied and basic biomedical science to human knowledge, health, and well-being (e.g., Strand and Strand, 1993). By the same token, animal rights activists often view scientists as cold-hearted technocrats who conduct trivial and cruel experiments for personal gain at the expense of helpless and innocent animals (e.g., Ruesch, 1978). Thus, scientists and animal activists seem to live in different intellectual worlds. The result has been a clash of paradigms that all too often makes misunderstanding inevitable and communication impossible (Gluck and Kubacki, 1991; Herzog, 1993a; Paul, 1995; Sutherland and Nash, 1994).

Scientists who have a reasonable grasp of the arguments raised by animal activists about the use of animals in behavioral and biomedical research and of the animal protection movement as a social phenomenon will be better able to discuss these issues with their colleagues, students, and the public. In this chapter, I review several aspects of the contemporary animal rights movement, including the moral philosophies underlying the movement and the results of recent investigations of the sociology and psychology of activism. I hope it will be of use to scientists who are seeking to understand animal activism as a social and psychological phenomenon and lead to better communication between parties who are caught in a bitter social conflict.

Philosophical Perspectives on the Moral Status of Animals: A Brief Introduction

Contemporary prescriptive ethical theories fall into one of two categories. *Utilitarian theories* are predicated on the belief that ethical decisions should be based primarily on the anticipated consequences of acts, the general aim being to maximize happiness and minimize suffering. *Deontological theories,* in contrast, hold that acts are right or wrong based on broad ethical principles (e.g., honesty, reciprocity, the possession of rights, respect for others, adherence to obligations) rather than positive or negative hedonistic outcomes. Neither set of theories provides unambiguous guidelines about how animals should be treated; that is, some utilitarians accord full moral status to animals, whereas others do not. Similarly, some deontologists argue that the same ethical principles apply to humans and animals, but others would have us give full moral consideration to our species alone.

The notion that animals and humans fall within the same moral sphere is of relatively recent origin. For most of the history of Western civilization, animals were not perceived to be within the realm of moral concern (Serpell, 1996). The ethical relationship between human and nonhuman animals was clear. Humans occupied a special moral niche, a position derived from the Judeo-Christian tradition in which God gives them dominion over other species. The classic proponent of this view was the seventeenth-century French philosopher René Descartes, who argued that animals did not merit moral concern as they did not have the capacity for language and free action. He considered animals to be mere biological automata whose behavior was governed by instincts and complex reflexes. Animals did not possess consciousness or a soul or feel pain. Thus, early anatomists apparently had no more moral compunction about dissecting a live unanesthetized dog than I might about ripping a bad memory chip from a balky computer (Phillips and Sechzer, 1989). (See Radner and Radner [1989] for an evaluation of Descartes's stance in the context of recent developments in ethology.)

The Cartesian position remained largely unchallenged until the nineteenth century. Ironically, given the hostility that some contemporary animal activists have toward science, it was a scientific breakthrough—Darwin's notion of organic evolution by natural selection—that provided the intellectual underpinnings for a revision in thinking about the moral standing of other species. The cornerstone of the evolutionary perspective is continuity between species in morphology and behavior. Indeed, animal research is predicated on the notion that some species are sufficiently like us that we can generalize information derived from animals to humans. But, similarity in physiology and behavior implies similarity of mental experience, such as the ability to experience pain and to suffer. Thus developed the inconvenient paradox: The more the use of a species is justified on scientific grounds, the more problematic is its use on moral grounds.

Darwin was aware of the ethical implications of his theory. In the first edition of *The Descent of Man and Selection in Relation to Sex* (1871) he wrote, "Every one has heard of the dog suffering under vivisection, who licked the hand of the operator: this man, unless he had a heart of stone, must have felt remorse to the last hour of his life." Darwin also believed that animal research was necessary for scientific progress; however, in the second edition of the book, he inserted into the sentence the proviso *unless the operation was fully justified by an increase in knowledge* (Burghardt and Herzog, 1980).

The Utilitarian Approach to Animal Liberation

The architect of the utilitarian approach to the treatment of animals was the eighteenth-century philosopher Jeremy Bentham. Bentham believed that ethical decisions should be based on a moral calculus that balances pleasure and pain. Bentham argued that it was not a creature's cognitive capacities that afforded it moral status but its capacity for sentience (i.e., the ability to experience pleasure and pain), hence Bentham's often quoted statement, "The question is not, Can they *reason,* nor Can they *talk,* but Can they *suffer?*"

The most prominent contemporary proponent of this view is the Australian philosopher Peter Singer. Singer's *Animal Liberation* (1975; revised 1990) is often referred to as the bible of the animal rights movement, a misnomer in that Singer's ethical stance is not based on the supposition that moral status hinges on the possession of rights. In *Animal Liberation,* Singer develops a coherent utilitarian argument for a revised ethic of human–animal interactions in an engaging and accessible style. The crux of Singer's argument lies in the *principle of equality:* All sentient creatures have an equal stake or interest in their own lives. Singer argues that this principle implies that there are no grounds for elevating the interests of one species, *Homo sapiens,* above all others. Discrimination against individuals on the basis of race or gender is morally wrong because the racist or sexist treats individuals differently based on morally irrelevant criteria. For Singer, the same is true of speciesism: "From an ethical point of view, we all stand on an equal footing—whether we stand on two feet, or four, or none at all" (Singer, 1975, p. 6). Like Bentham, Singer argues that the only relevant criterion for moral consideration is the capacity to suffer. By definition, sentient animals are capable of suffering and, therefore, deserve moral concern.

Where should we draw the line between sentient and nonsentient beings? In the first edition of *Animal Liberation,* Singer drew it "somewhere between the shrimp and the oyster," although he later confessed to being uneasy with this demarcation (Singer, 1990). Note that Singer does not claim that all animals must be treated alike, only that their interests must be duly considered and that they not be discriminated against on the basis of species.

According to Singer, the use of animals in scientific experimentation is a prime example of speciesism and is directly analogous to the racist use of humans in "medical" experiments in Nazi Germany. In theory, Singer would permit the occasional use of animals in research under circumstances in which a single experiment would save many lives. Furthermore, to avoid the charge of speciesism, the experiment must be so important that we should also be prepared to carry out the study on a human child with mental abilities equivalent to that of the animal subject. In practice, Singer believes that these conditions are rarely met by scientific research.

The "Rights" Approach

Singer's utilitarian logic is not the only intellectual route to a revised perspective on how animals should be treated. An alternative is found in the deontological argument that animals are entitled to basic rights. This argument is most forcefully made by Tom Regan (1983) in his treatise *The Case for Animal Rights*. All rights proponents hold that some creatures are entitled to fundamental rights such as life, bodily integrity, and basic moral consideration. This being said, questions inevitably arise as to which creatures are entitled to rights and on what grounds. Traditional ethical systems are anthropocentric. Rights holders are restricted to beings that meet certain minimal cognitive criteria that have historically been considered uniquely human, including the capacity for language, self-consciousness, the ability to enter into reciprocal contractual obligations, a sense of right and wrong, and membership in a moral community.

Animal rights theorists expand the criteria to include some nonhuman animals. For Regan, the fundamental criterion for possessing rights is *inherent value*. To have inherent value, Regan requires that an animal possess more than just the capacity for pain and pleasure. They must be capable of higher level mental abilities such as beliefs and desires, emotion, memory, intentions, a sense of the future, and a "psychological identity." Regan notes that the actual phylogenetic line separating species that possess these attributes from those that do not is currently unclear and should, for the time being, be "drawn in pencil." Nonetheless, he holds that these criteria are at least met by mammals one year of age and older. Animals with these capacities are the *subjects of a life*. They possess inherent value and, consequently, moral rights.

Central to Regan's view is the idea that inherent value is an all-or-none trait; all creatures that have inherent value have it in equal measure. Human life has no more ethical value per se than that of a pig. Like humans, animals are entitled to certain fundamental rights including the right to be treated with respect and the right not to be harmed. As with Singer's ethical theory, the rights view does not mean that all animals should be treated the same way. Animal rights advocates, for example, do not argue that dogs should be given the right to drive or to vote.

To Regan, the scientific use of animals is immoral because researchers treat animals as renewable resources—as objects rather than as subjects of a life. In addition, he does not believe that it is acceptable to sacrifice an innocent few for the benefit of the many. Thus, Regan is a total abolitionist when it comes to animal experimentation. In his view, there is no moral justification for any research that might harm an animal subject. His agenda for the animal rights movement is clear: "That movement, as I conceive it, is committed to the following goals, including: total abolition of the use of animals in science; the total dissolution of commercial animal agriculture; the total elimination of commercial and sport hunting and trapping" (Regan, 1987, p. 46).

Comparisons

The two most prominent animal liberation philosophers do not entirely agree on why humans' exploitation of animals is morally wrong. Indeed, Singer points out the philosophical problems with rights-based arguments for animal liberation, and Regan does the same for utilitarian ethical thinking. There are, however, important commonalities in their positions. While acknowledging the differences between humans and animals, both feel that these differences are not relevant to the issue of whether a creature is entitled to basic moral consideration. Both deplore speciesism. Further, Regan and Singer largely come to the same conclusions about the morality of current uses of animals, even though they take different paths to get there. The logical extension of their arguments is the elimination of customs that involve the enforced captivity or killing of animals. Practices such as eating animals, using them for research, caging them in zoos, and trapping them for their fur are equally wrong under both positions.

Although Singer's utilitarian and Regan's rights arguments are the best known systematic philosophical arguments against the exploitation of nonhuman animals, other approaches have been offered. See, for example, the work of Adams (1990) for a feminist perspective on animal rights, Linzey (1987) for a Christian view, and Rachels (1990) for a Darwinian approach.

The Psychology of Animal Activism

Regan, Singer, and other moral philosophers have developed coherent ethical theories that propose fundamental revisions in the relationships humans have with members of other species. It is unlikely, however, that the development of the contemporary animal rights movement, beginning in the 1970s and continuing into the 1990s, was simply the result of the publication of books such as *Animal Liberation, The Case for Animal Rights,* and the numerous other scholarly publications by animal protection philosophers (e.g., Midgley, 1983; Rachels, 1990; Rollin, 1981; Sapontzis, 1987) or their critics (e.g., Carruthers, 1992; Cohen, 1986; Frey, 1980; Leahy, 1991). Clearly, a substantial segment of the public was predisposed to receive this message. Indeed, the philosophical debate over the moral status of animals provides little insight into the development of animal activism as a social and psychological phenomenon. Individuals seeking to understand animal liberation as philosophy and animal activists as people are engaged in somewhat different endeavors.

Demographics

Several approaches have been taken by social scientists interested in who becomes an animal activist and why. Some researchers have used traditional quantitative research techniques such as surveys and questionnaires to characterize the typical animal activist based on demographic characteristics. The samples in these studies have been drawn from animal rights demonstrators who attended large national animal rights demonstrations (Galvin and Herzog, 1992; Galvin and Herzog, in press; Jamison and Lunch, 1992; Plous, 1991; Plous, in press) and from subscribers to a major animal rights magazine (Richards and Krannich, 1991).

The data gathered by Shelley Galvin and me on animal rights demonstrators (Galvin and Herzog, 1992; Herzog, 1995) are typical of findings on demographic aspects of the animal rights movement. We distributed surveys to 600 individuals who were participating in the 1990 March for the Animals in Washington, D.C. The march was the largest animal rights demonstration yet held in the United States, attracting a crowd estimated at 20,000 to 30,000 activists. We asked the participants to fill out the survey after the demonstration and return it to us via a preaddressed postage-paid envelope. While the survey contained a variety of demographic and attitude questions, we were particularly interested in the philosophical basis of moral judgment of animal activists.

Activists returned 157 questionnaires. The sample comprised 121 women and 36 men. The respondents ranged in age from 14 to 71, with a median age of 32. They had been involved in animal rights issues for a median of three years (range = 1 to 30 years). Of the respondents, 95 percent were white; there were no blacks. Forty-six percent were single, and 54 percent were married. The majority (77 percent) were from urban areas or suburbs. As a group, the activists were well educated; 78 percent had pursued education beyond high school, with 49 percent completing at least four years of college. About half listed their occupations as professional, 17 percent were skilled, 7 percent were semiskilled or unskilled, 19 percent were students, 3 percent were retired, 3 percent were homemakers, and 1 percent listed their occupation as professional animal rights activist. Only 30 percent were members of mainstream religious denominations, and about half claimed to have no religious preference or were atheists or agnostics.

Lifestyle Changes. Involvement in the animal rights movement inevitably leads to changes in behavior and lifestyle. Among our sample, the most pronounced area of lifestyle change was diet; 97 percent of the respondents reported having significantly changed their eating habits, and 75 percent were either vegan (individuals who consume no animal products at all) or vegetarian. Participation in the movement was also manifested by a variety of other behaviors including the purchase of "cruelty-free" consumer products (4 percent), boycotting companies that tested products on animals (93 percent), donating money to animal rights organizations (93 percent), wearing animal-free clothing (84 percent), displaying animal rights bumper stickers or wearing animal rights buttons (79 percent), writing letters to legislators (75 percent), and participating in protest rallies and marches other than the March for the Animals (65 percent).

Animal Rights Activism and Moral Orientation. One of the most salient characteristics of animal activists is that they take moral issues very seriously. For many, the cause of animals becomes a dominant theme in their lives and affects many aspects of their daily routine. This fact suggests that animal activists have a different approach to moral thinking than the majority of Americans. Galvin and I tested this idea by including in our research packet the ethics position questionnaire (EPQ). This instrument was developed by Forsyth (1980) to assess individual differences in personal moral philosophy. The EPQ measures two dimensions of moral judgment: ethical relativism and ethical idealism. According to Forsyth, relativism is the degree to which individuals reject the belief that ethical decisions should be based on absolute and universal moral principles. Idealism refers to the degree to which individuals believe that moral behavior inevitably results in positive outcomes.

The EPQ consists of 20 Likert scale statements, half designed to assess relativism and half designed to measure idealism. Ethical orientation is determined by median splits of the two dimensions. This process results in a two-by-two matrix, with each cell representing individuals with different ethical ideologies:

> *Situationists* (high relativism, high idealism) believe there are no universal moral principles to guide behavior and that well-being is maximized and interests are protected through cost-benefit analyses of individual acts.
>
> *Absolutists* (low relativism, high idealism) hold that there are universal moral principles and that adherence to them will generally lead to positive consequences and the protection of the general welfare.
>
> *Subjectivists* (high relativism, low idealism) do not believe that morality should be based on universal principles or that doing the right thing will always produce positive outcomes; they view personal values as the ultimate basis for moral judgment.
>
> *Exceptionists* (low relativism, low idealism) subscribe to universal moral principles, yet they are pessimistic about the prospect that good results inevitably result from moral behavior. As a result, they are pragmatically open to exceptions to universal principles.

The EPQ has been found to have good psychometric properties and is related to attitudes toward a variety of social issues (Forsyth, 1980; Forsyth, Nye, and Kelly,

1987). For comparative purposes, we also administered the EPQ to a group of 198 college students from Western Carolina University. Our hypothesis was that animal rights activists would be more absolutist and less situationist in their moral orientation than the comparison group.

As shown in Figure 11.1, the animal rights demonstrators had a different distribution of ethical ideologies than the college students. Three-fourths of activists were classified as absolutists in moral orientation, as were 25 percent of the students. Furthermore almost 20 percent of the students were subjectivists, but none of the demonstrators fell into this category.

The majority of the activists who returned our questionnaire were committed to an ethical ideology that is idealistic and nonrelativistic. They were moral optimists. (See Galvin and Herzog, in press, for a discussion of the relationship between animal activism and psychological optimism.) This finding, if characteristic of the movement as a whole, has several implications. First, absolutists may be less willing to compromise than individuals who view ethics through a more relativistic lens. Hence, it is not surprising that scientists and activists have difficulty in reaching common grounds for discussion. Second, moral orientation may be a factor that predisposes some individuals toward becoming involved in a cause such as the animal rights movement and virtually precludes this type of commitment in others. In this

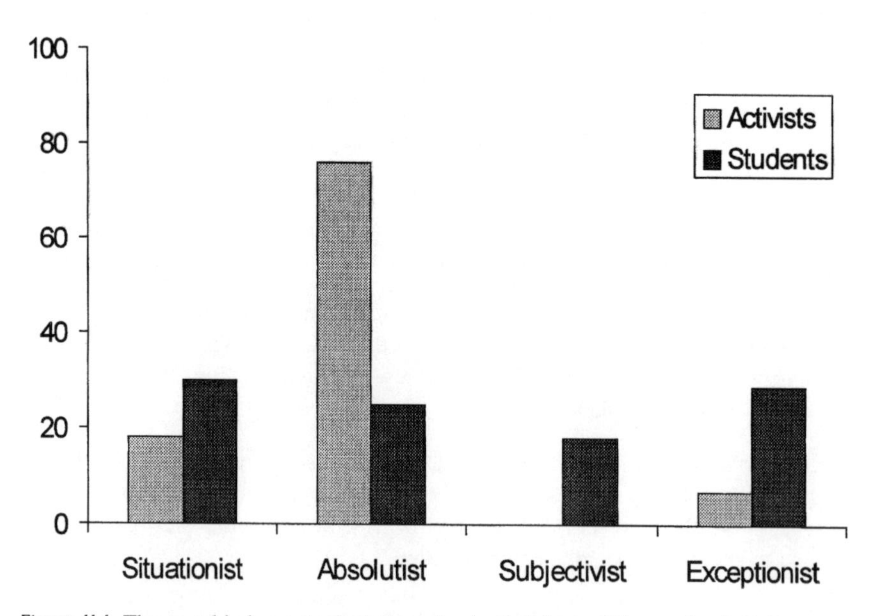

Figure 11.1 The moral judgment orientation of animal rights activists and university students in terms of ethics position questionnaire categories.

light, we were particularly interested in the finding that there were no activists in the subjectivist category (high relativism, low idealism), the most cynical of the four moral orientations.

Several provisos should be kept in mind to interpret these results. Our return rate was slightly less than 30 percent. While this rate is typical of this sort of mail-in survey, it is possible that our sample was not representative of participants in the march. There is, however, evidence to suggest that these results do reflect the attitudes and beliefs of the demonstrators. Two other research teams were also gathering data at the march. Jamison and Lunch (1992) and Plous (1991) obtained survey and interview data on randomly selected march participants. Because their data were gathered on the spot, they were not subject to the response bias of our mail-in survey. (We used the mail-in method because of the length of our questionnaire.) However, the pattern of demographic data that we gathered was virtually identical to the results of the other researchers. The fact that our data were consistent with data obtained using different methods suggests that they are reasonably representative of the attitudes of the participants.

The use of individuals attending a national demonstration to represent animal activists as a group is also a limitation. Activists who have the time, desire, and financial means to travel to a national demonstration may be more committed, absolutist, and affluent than the population of animal rights activists as a whole. Although this possibility cannot be precluded, our findings were consistent with information gathered on animal activists who were not attending demonstrations. Richards and Krannich (1991) surveyed a sample of subscribers to *The Animal's Agenda,* a prominent animal rights magazine. Their demographic data also closely paralleled our findings.

The Lives of Animal Activists: A Narrative Approach

Paper-and-pencil surveys and closed-ended questionnaires can provide useful information on aspects of a social movement, such as its demographic base and the attitudes of the participants. Quantitative research techniques, however, are not particularly good at exploring other dimensions of human lives. To investigate more complex aspects of the psychology of animal activism, researchers have turned to qualitative methods such as participant observation and the analysis of interviews and electronic communications (Groves, 1997; Herzog, Dinoff, and Page, 1977; Jasper and Nelkin, 1992; Shapiro, 1994; Sperling, 1988). These methods offer nontraditional alternatives to standard research techniques in the social sciences. Typically, qualitative research involves observation and interviews with a relatively small group of individuals chosen for their experience rather than by random selection (Bogdan and Biklen, 1992; Lincoln and Guba, 1985; Patton, 1990). Investigators who use this approach are more concerned with insight than statistical rigor, and the qualitative approach is not as scientific in the strict sense; however, it can offer considerable insight into human experience. (See Arluke and Sanders [1996] for examples of the application of qualitative methods to the study of human–animal interactions.)

Over a three-year period, I used qualitative techniques to investigate the lives of several dozen animal activists. I attended meetings and demonstrations and conducted extensive interviews with the participants. The interviews were tape-recorded and transcribed. The questions focused on how the activists became involved with the animal rights issue, how their lives were affected by their involvement, issues associated with consistency between belief and behavior, perceived moral "gray areas," and personal benefits and costs of involvement in the movement (see Herzog [1993b] for a detailed description of the interview methods).

I was interested in the psychology of rank-and-file animal activists—individuals who write letters to congressional representatives, march at demonstrations, make changes in their lives because of their beliefs, and think of themselves as animal rights activists. Jasper and Nelkin (1992) classified animal protectionists into three categories: welfarists, pragmatists, and fundamentalists. The participants in my study were generally fundamentalists—people who hold that individuals do not have the right to "use animals for their own pleasures or interests, regardless of the benefits" (Jasper and Nelkin, 1992, p. 9). I did not interview people whose major orientation was animal welfare, such as members of local humane societies or individuals who expressed sympathy for the plight of animals but who did not make the lifestyle changes that more committed individuals made. Nor did I interview nationally prominent spokespeople who might be expected to produce rehearsed responses to my questions. Only a small minority of animal rights proponents are involved with clandestine organizations, such as the Animal Liberation Front, which advocate and occasionally conduct illegal or violent activities, and none of the participants in my study reported involvement in these activities. All but two of the participants resided in the southeastern United States at the time of the interviews, and some of their answers may reflect this regional distribution. Despite these provisos, I believe that the people I interviewed are fairly typical of individuals who attend fur protests, display animal protection bumper stickers, and express their convictions to coworkers, family members, and friends. There were 7 males and 16 females in my sample, a sex ratio that roughly mirrors the gender composition of the movement. The activists ranged in age from 14 to 71, and they had been involved in the movement from 1 to 10 years.

The Search for Moral Consistency

I asked the participants about the relationships between their lifestyles and their beliefs. All the individuals I interviewed were striving to achieve consistency between their behavior and their ideals. Perhaps the most fundamental issue all of them had to deal with was diet. All 23 were vegetarians, though to different degrees. Fourteen were either complete or aspiring vegans (individuals who consume no animal products, including milk, eggs, and honey). Several others, although they considered themselves vegetarians, ate eggs and, in one case, fish.

Among the participants, there was a complex interaction between diet and thinking about the treatment of animals. About half of the individuals were vegetarians for health reasons before they took on the mantle of animal rights. Others came to perceive vegetarianism as a moral imperative only after they had made a shift in their ethical thinking. In some cases, the change to a meatless diet came easily; for others, it was more difficult. A few admitted to occasionally longing for a smoked sausage even after years of vegetarianism. Diet and ethical thinking were sometimes synergistically related. As one women said, "The more I got involved, the more my diet changed. And the more my diet changed, the more I got involved."

There were other manifestations among the activists of the attempt to achieve moral consistency. Most bought consumer products, such as cosmetics, that were not developed through animal testing. One woman had been forced to declare bankruptcy because of the money she and her husband had put into animal rights organizations. Another explained why she refused to use flea-and-tick powder on her dog. Instead she picked them off the dog one by one and released them outside her home. She said of the fleas and ticks, "I know they do not feel pain or anything, but I feel it is important to be consistent. If I draw the line somewhere between fish and mollusks, it isn't going to make sense."

This search for moral consistency became a major dynamic in the lives of some of the activists, giving them a source of moral satisfaction. Dave, an accountant, told me:

> On a personal level, after two years of vegetarianism, I can honestly say I feel good about knowing that I can go through my life—my entire day—without being responsible for any cruelty. I don't wear any animal products, all my toiletries are cruelty-free. My household products are cruelty free. Personally, it gives me a feeling of freedom. I am free. In exploiter-and-the-exploited relationships, the exploiter is held prisoner in a sense. I am no longer an exploiter.

However, the burden of living up to high moral expectations sometimes proved difficult. For example, one of the activists, an avid softball player, had become concerned with the moral implications of his sport. After a long search, he had obtained a plastic softball glove, but he had been unable to locate a source of balls that were not encased in leather. He was considering giving up a pastime that gave him great pleasure.

Pets were also a source of moral concern. Cats were a particular problem as they do not thrive on the vegetarian diets that their human guardians opted for. They were not the only animals that posed moral problems. One man, a graduate student in statistics, relayed a story about his parrot:

> Animals are not here for our happiness. Up until recently I had a parrot. I would let him fly around my room. One day I just looked at him and said to myself, "This is wrong. It wants to be free." I just took it out in the back yard and let it go, even though it was hand-fed. And I knew it wouldn't survive in the wild. It was great, amazing. I was really happy to see it fly. I assume that he probably starved to death. It may have been something that I was doing for myself rather than for the bird.

Cognitive Aspects of Animal Activism

Thoughts about the treatment of animals assumed a dominant role in the day-to-day mental lives of many of the activists. When I asked one man, a doctoral candidate in zoology, how often he thought about the treatment of animals, he said, "Absolutely all the time. I'd say every waking moment, if not every second." Another, an undergraduate journalism major, responded, "It is always on my mind. It keeps me up at night. I don't care where I am or what I'm doing, something always pops into my mind about the movement. For example, a picture I've seen of a cat in a stereotaxic device or something. It's just always there. I constantly think about it."

Few of the activists had more than a passing knowledge of the philosophical basis of the movement. Although about half had read *Animal Liberation,* only a few were able to articulate the nuances of animal rights philosophy, such as the debate between the rights theorists like Regan and utilitarians like Singer. For the most part, they found these arguments academic and of little personal interest; however, four of the activists were well versed in the philosophical arguments for and against animal rights.

The animal rights opponents often accuse activists of being overly emotional. I asked the participants about the relative role of reason and emotion in their involvement in the movement. For some, the appeal of the movement was almost entirely emotional. One woman, a special education teacher, told me, "A lot of this has an emotional basis for me. There is a literal pain—the kind of pain that you might feel about hearing that someone has died. I feel that way when I hear about animals that are suffering in laboratories." Another admitted that she was initially drawn to the movement because of an empathy with animal suffering. She went on, however, to acknowledge the importance of intellect in her involvement: "It is becoming more intellectual for me because you can't argue with people on an emotional basis. Nobody respects you if you just say 'I love dogs and cats.' So you have to get to where you can explain the theory behind why you do it. Without logical reasons no one is going to pay attention."

For some activists, the balance between emotion and reason tilted toward the intellectual. One women resented being referred to by friends as softhearted: "I don't like the term 'softhearted.' It makes it an emotional issue . . . I think that to pass off all of the years that I have been thinking about these issues as being softhearted is really condescending to me."

An extreme example of intellectually based commitment was a man who had once worked in a poultry research laboratory. One aspect of his job was to dispatch research animals after they had been used for data collection. Some days he had to kill as many as 300 baby chicks, which he did, in his words, "by breaking their necks." He told me that this procedure did not bother him at all: "I don't think of it as an emotional issue. I think I could still go back into the laboratory and do all of those horrible things to animals. When I see things happen to animals it does not affect me viscerally."

Even the most vociferous animal rights proponent comes across animal treatment issues and situations in which moral judgments are difficult. I asked the activists what they considered to be ethical gray areas—uses of animals in which their thinking was uncertain or unclear. In this context, they mentioned keeping pets, taking drugs that had been tested on animals, the use of animals in medical research, wearing leather products they had acquired before joining the movement, and killing insects. A few had even considered the ethical implications of eating certain types of plants, asking, for example, "Is it more acceptable to eat fruit or nuts rather than, say, a carrot because the latter dies when it is consumed?"

The ethics of civil disobedience was also a moral gray area. Most of the individuals I interviewed believed that it was counterproductive—a bad strategy. A few others, however, felt it was necessary to achieve the objectives of the movement. All but two opposed the use of violence against individuals, such as researchers whom they perceived as animal exploiters. A corporate executive with two young children expressed her grudging admiration for the more radical branch of the movement: "I am glad that there are people who are able to take violent action out there. But, I'm also glad that I'm not one of them. I'm glad there is someone else to do the dirty work."

Attitudes Toward Science

It is no surprise that the activists I interviewed generally opposed the use of animals in science. (See Herzog, Dinoff, and Page (1997) for a description of extended debates that took place between scientists and animal activists over an electronic bulletin board.) Their opposition was on both practical and moral grounds. Scientists argue that animal research has resulted in major gains in human health and well-being in this century. For example, a policy statement of the American Medical Association (1992) states: "Virtually every advance in medical science in the twentieth century, from antibiotics and vaccines to anti-depressant drugs and organ transplants, has been achieved either directly or indirectly through the use of animals in laboratory experiments" (p. 11).

Many of the activists I interviewed vociferously disagreed with this view. On the contrary, some of the activists felt that animal research has actually inhibited biomedical research. When asked if she felt that medical progress would continue at its present rate if animal research were discontinued, one of the participants replied:

> Oh, definitely. I don't think that they will ever find a cure for AIDS or cancer unless scientists get off animal research. Yes, animals get cancer, but it is different from our cancer. Take cigarette smoke. They force smoke on all these animals. They do it to rats and monkeys. They never get lung cancer, but we do. The main thing that kept us from finding a good anesthesia for so long was that they kept testing it on dogs. Morphine excites dogs instead of knocking them out like it does us.

Another said:

> I can't find any substantial data to indicate that anything has been discovered from animals that could not have been gathered from some other source. Thalidomide was a drug that was found to be safe on experimental animals and given to the human population. It was a complete and total disaster. They say that vaccines were discovered from animal research. But, the Chinese had vaccines before the time of Christ, and they were not found from animal research. They were discovered from actual clinical trials and human patients.

In several cases, activists took a blame-the-victim stance toward individuals suffering from chronic disease, particularly illnesses associated with lifestyle and diet. I was frequently told that if people would just give up eating meat, diseases such as cancer and heart disease would disappear. For example, a university student said, "People are ill because they have made themselves ill—whether it is that person or something that happened three generations ago. It is people screwing up the environment or smoking or overeating. These are human problems, not animal problems."

One activist felt that humans should be substituted for animals in experiments: "I have a thing about death row inmates. I really think that they should be used for lab experiments. I wish I had a bumper sticker that said 'Liberate Lab Animals—Use Death Row Inmates Instead.' "

Another, however, used the death row–inmate scenario to argue against animal research on loftier grounds:

> Why don't we go and use prisoners off death row? They would be the best subjects for medical research. We could cut them up at the end of the experiment and see how the tests worked. We don't do it. And I agree that we shouldn't. Even if he were a rapist and a murderer and is going to die next Wednesday, we must respect his inherent right not to be used as a tool. We put this ethical curb on ourselves and thereby forgo benefits that would be achieved by using prisoners as research subjects. But, why should we stop [applying these ethical principles] at the line of other species?

Animal research was also seen as trivial and repetitious by the activists. Typical was the comment by a woman who told me, "There is a lot of repetition in these experiments. How many times do you need to feed rabbits Drano to know it is going to die."

Some activists viewed scientists as committing a kind of hubris that was overturning the natural order. For example, one activist said, "I sometimes wonder if we are trying to control too many things. Maybe that child is supposed to die, horrible as that sounds. Maybe it is the survival of the fittest." Others were quite cynical about the motivations of scientists. One said:

> This research is all due to the buddy-buddy and good ole boy system. Research grants are doled out by the same people who are going to be asking for grants the next year. If I am on the grant board and you want $50,000 to tie cats to a plank in a pool of water to see how long they will stay awake before they give up and fall over, and I say, "Well, this is silly and cruel," and I am the deciding vote, you don't get the money. Then the next year I am up for a grant and you are in the same position. I wonder if you are going to look at the merits of my experiment and say, "Well, that's the SOB that wouldn't let me get my money last time."

She went on to say that cancer research has "made more millionaires than any cause in history."

Most of the participants felt that scientists themselves were not inherently evil, but one man, an accountant, expressed his rage at animal researchers:

> It doesn't just piss me off. I get really angry. Animals are subjected to this ridiculous torture, and they don't know what is going on. They cannot cope, and they do not know what is being done to them. They are in laboratories, and needles are being stuck into them. Their necks are broken, and they are held in these stereotaxic devices for days or a week without even moving. They are just terrorized. The more you think about it, the more offensive it is because it is just incredibly sadistic. Scientists are specifically dreaming up deviant ways of torture, not unlike concentration camps.

Relations with Other People

Involvement in the animal rights movement had affected the interpersonal relationships of many of the activists. While most of the respondents reported that their family and friends were supportive of their activities, the intense psychological involvement, behavior changes, and philosophical differences associated with the animal rights issue played a significant role in the breakup of two marriages, with a third faltering at the time of the interview. Husbands and wives sometimes made uneasy compromises. One woman, a psychologist, described with pride the accommodations she had made with her husband, who was an animal researcher. She told me that she had agreed even to cook meat for him. As an afterthought she added, "But, of course, I would never kiss him after he has been eating meat."

In contrast, the cause of animals sometimes became a family commitment. In four cases, husbands and wives were both dedicated to the movement, and it was source of shared meaning in their relationships.

The activists commonly spoke of the effect their commitment had on their social lives. Typical were the comments of a college professor:

> It is really hard for me to make friends these days because I really don't like going out to eat. There are a lot of awkward social situations. At work, I avoid eating lunch with people because I just don't like being around it. It has had a toll on my relationships with family, friends, so on. It is kind of a strain on interpersonal relations, that's for sure.

There was an evangelical component to the activism of the participants. Almost all of them expressed the need to spread their message. I was frequently told statements to the effect, "If everyone knew what I know about factory farms, animal research, etc., they would change their lives as I have mine." For most, the ardor of the activists took the form of modeling, setting an example rather than browbeating. Nevertheless, one respondent delighted in surreptitiously placing stickers that said "Warning: This Package Contains Dead Animals" on packages of meat in her local supermarket.

Feelings of Moral Superiority

Most of the participants seemed to have developed a sense of moral superiority that they felt separated them from others who did not share their vision or understand their convictions. I do not mean this in a negative sense. Like religious converts, the activists had discovered personal truth. When I asked one activist about this aspect of her life, she responded:

> Yes, there is the sense that I'm not part of the system that keeps animals in the food chain. It is freeing—almost like that feeling that you have when you've just showered—that squeaky clean feeling. But then sometimes I come home from work and find myself yelling at the kids, and then I don't feel so good anymore. That squeaky feeling doesn't stay with me all the time.

Not all the participants, however, felt this way. Five denied having any sense that they felt morally superior to those with other lifestyles.

Personal Happiness

For most of the activists, their involvement with animal rights involved a life-changing experience. Certain satisfactions follow from the acceptance of a new moral paradigm. As one activist put it, "This gives me something to live for. I feel sorry for other people who do not have this type of thing in their lives." For others, though, the conversion to an animal rights perspective had some negative consequences that ultimately led to considerable unhappiness. They bore a significant moral and psychological burden. Some felt guilty when they could not live up to their convictions; the simple act of swatting a mosquito or putting flea powder on a dog took on moral dimensions. Several mentioned that they could not sleep at nights sometimes when their thoughts became haunted by images of animal suffering.

This burden sometimes seemed overwhelming. In the latter interviews, I asked if the costs of involvement ever seemed too great. This question struck a chord within most of the interviewees. One activist said:

> I am burning out. After five years I have come to the point of near emotional collapse. My life is so full of the movement now that I have no spare time anymore. I have thrived on this in past years but suddenly this Easter it came to a point where I said, I just can't do it anymore. I don't have the strength anymore.

In one of the more poignant interviews, a women told me:

> I don't think that most people really think I'm nuts. But, I think sometimes that I'm nuts, because I drive myself crazy about it. It dominates my life. Sometimes I think I can't take it any more. I can't think about this 24 hours a day. So I just say I'm going to back off a bit. I'm going to loosen the rope a little. I'm going to let myself not be Jesus for a minute and be a normal human being.

Animal Activism as Religion

Jasper and Nelkin (1992) refer to the animal rights movement as a "moral crusade." As is often the case with crusades, there is a religious flavor to the animal rights movement (Sperling, 1988). Animal activists are not necessarily religious in a conventional sense. Indeed, more of our March for the Animals respondents claimed to be atheists or agnostics than were members of mainstream denominations. However, there are significant similarities between religious conversion and the assumption of an animal rights perspective. These similarities include commitment to a moral ideal, sacrifice, a change in lifestyle, evangelism, and even a sense of sin. As one of the individuals I interviewed said, "It is my religion. Certainly there is no church involved. But, sure it is religious. Religion is the way you live your life. It is a spiritual thing not to want to kill. It is a religious lifestyle because you are surrounded by it." Another individual told me:

> There hasn't been much of a religious aspect to my activism in animal rights issues, but I have grown to respect Jesus in a very different way. Most people who worship Jesus are very conservative and conformist. But Jesus was an activist. I think that if Jesus were alive today certainly he would be a vegetarian, and I think he would be an animal rights activist.

Conclusions

Scientists who make the attempt at dialogue with animal activists need to recognize that not all animal activists are alike. Some are drawn to the movement primarily by their emotional response to animal suffering; others are attracted by the logic of moral philosophy. For some, the conversion to an animal rights perspective occurs via a gradual series of stages. For others, the change occurs literally overnight. Although many activists describe an early sensitivity toward the mistreatment of animals (Shapiro, 1994), others do not. Animal activists are not in agreement over the philosophical basis of animal liberation or unanimous in their beliefs about specific uses of animals.

Nonetheless, some general themes emerge from studies of the psychology of animal activism. For many of those involved, the elimination of animal suffering at human hands has become a major focus of their lives. It affects what they eat and wear, what they think, and who their friends are. There are psychological benefits and costs to this type of involvement.

At present, the question of why some people and not others with the same information are drawn to the animal rights movement has no clear answer. I suspect that some individuals are predisposed toward animal activism by a combination of early empathy with animals (a common theme in my interviews), an absolutist moral worldview, and an optimism that individuals can make a difference. More research, however, is needed to answer questions related to the psychological underpinnings of animal activism. For example, what are the similarities and differences between animal advocates and other social activists such as antiabortionists and environmentalists?

Whatever the causes of their convictions, animal rights activists live in a different moral universe from most people. This difference in paradigm makes it hard for scientists and activists to communicate. Like debates between evolutionists and creationists, the protagonists do not share the same basic assumptions. Too often, the parties become intransigent, and communication becomes difficult, if not impossible. As one participant told me, "I do have the sense that what I am doing is right, and if you argue with me I'm not going to listen because I know I am right." However, there is something commendable in the moral commitment of animal activists. They have the courage to try to bring their personal behavior in line with their convictions and the idealism to try to change the world. But there are negative aspects to this type of moral commitment. Occasionally, it can result in rigidity, fanaticism, and violence. However, in a world in which so many seem to lack any kind of moral compass, there may be something admirable about marching to the beat of a different drummer.

References

Adams, C. *The Sexual Politics of Meat*. New York: Continuum, 1990.

American Medical Association. *Use of Animals in Biomedical Research: The Challenge and Response.* Chicago: American Medical Association, 1992.

Arluke, A. and C. Sanders. *Regarding Animals.* Philadelphia: Temple University Press, 1996.

Bekoff, M. Experimentally induced infanticide: The removal of birds and its ramifications. *Auk* 110: 404–406, 1993.

Bogdan, R., and S. K. Biklen. *Qualitative Research for Education*. Boston: Allyn and Bacon, 1992.

Burghardt, G. M., and H. A. Herzog Jr. Beyond conspecifics: Is Brer Rabbit our brother? *Bioscience* 30: 763–768, 1980.

Carruthers, P. *The Animals Issue: Moral Theory into Practice*. New York: Cambridge University Press, 1992.

Cohen, C. The case for the use of animals in biomedical research. *N. Engl. J. Med.* 315: 865–869, 1986.

Darwin, C. *The Descent of Man and Selection in Relation to Sex.* London: Murray, 1871.

Emlen, S. T. Ethics and experimentation: Hard choices for the field ornithologist. *Auk* 110: 406–409, 1993.

Forsyth, D. R. A taxonomy of ethical ideologies. *J. Personal. Soc. Psychol.* 39: 175–184, 1980.

Forsyth, D. R., J. L. Nye, and K. Kelly. Idealism, relativism and the ethic of caring. *J. Psychol.* 122: 243–248, 1987.

Frey, R. G. *Interests and Rights: The Case Against Animals*. Oxford: Clarendon Press, 1980.

Galvin, S., and H. A. Herzog Jr. Ethical ideology, animal activism and attitudes toward the treatment of animals. *Ethics Behav.* 2: 141–149, 1992.

Galvin, S. L. And H. A. Herzog Jr. Attitudes and dispositional optimism of animal rights demonstrators. *Society and Animals,* in press.

Gluck, J. P., and S. R. Kubacki. Animals in biomedical research: The undermining effect of the rhetoric of the besieged. *Ethics Behav.* 1: 157–173, 1991.

Groves, J. *Hearts and Minds: The Controversy over Laboratory Animals.* Philadelphia: Temple University Press, 1997.

Herzog, H. A. Human morality and animal research. *Am. Scholar* 62: 337–349, 1993a.

Herzog, H. A. "The movement is my life": The psychology of animal rights activism. *J. Soc. Issues* 46: 103–119, 1993b.

Herzog, H. A., Jr. The psychology of animal rights activism. In A. M. Goldberg and L. F. M.

van Zutphen, eds., *Proceedings of the World Congress on Alternatives and Animal Use in the Life Sciences: Education, Research, Testing*. New York: Mary Ann Liebert Publishers, 1995.

Herzog, H. A. Jr., B. Dinoff and J. R. Page. Animal rights talk: Moral debate over the Internet. *Qualitative Sociology* 20: 399–418, 1997.

Jamison, W., and W. Lunch. The rights of animals, science policy and political activism. *Sci. Technol. Hum. Values* 17: 438–458, 1992.

Jasper, J. M., and D. Nelkin. *The Animal Rights Crusade: The Growth of a Moral Protest.* New York: Free Press, 1992.

Leahy, M. *Against Liberation: Putting Animals into Perspective.* London: Routledge, 1991.

Lincoln, Y., and E. G. Guba. *Naturalistic Enquiry.* Beverly Hills, Calif.: Sage, 1985.

Linzey, A. *Christianity and the Rights of Animals.* New York: Crossroads, 1987.

Midgley, M. *Animals and Why They Matter*. Athens: University of Georgia Press, 1983.

Orlans, F. B. *In the Name of Science: Issues in Responsible Animal Experimentation*. New York: Oxford University Press, 1993.

Patton, M. Q. *Qualitative Evaluation and Research Methods*. Newbury Park, Calif.: Sage, 1990.

Paul, E. Us and them: Scientists' and animal rights campaigners' views of the animal experimentation debate. *Soc. Anim.* 3: 1–21, 1995.

Phillips, M. T., and J. A. Sechzer. *Animal Research and Ethical Conflict*. New York: Springer-Verlag, 1989.

Plous, S. An attitude survey of animal rights activists. *Psychol. Sci.* 2: 194–196, 1991.

Plous, S. Signs of change within the animal rights movement: A follow-up survey of activists. *Journal of Comparative Psychology,* in press.

Rachels, J. *Created from Animals: The Moral Implications of Darwinism.* Oxford: Oxford University Press, 1990.

Radner, D., and M. Radner. *Animal Consciousness*. Buffalo, N.Y.: Prometheus, 1989.

Regan, T. *The Case for Animal Rights.* Berkeley: University of California Press, 1983.

Regan, T. *The Struggle for Animal Rights*. Clarks Summit, Pa.: International Society for Animal Rights, 1987.

Richards, R. T., and R. S. Krannich. The ideology of the animal rights movement and activists' attitudes towards wildlife. *Trans. North Am. Wildl. Nat. Res. Conf.* 56: 363–371, 1991.

Rollin, B. E. *Animal Rights and Human Morality.* Buffalo, N.Y.: Prometheus, 1981.

Rollin, B. E. *The Unheeded Cry.* Oxford: Oxford University Press, 1989.

Ruesch, H. *Slaughter of the Innocent.* New York: Bantam Books, 1978.

Sapontzis, S. F. *Morals, Reason and Animals*. Philadelphia: Temple University Press, 1987.

Serpell, J. *In the Company of Animals: A Story of Human-Animal Relationships* (rev. edition). Cambridge, UK: Cambridge University Press, 1996.

Shapiro, K. The caring sleuth: Portrait of an animal rights activist. *Soc. Anim.* 2: 145–165, 1994.

Shapiro, K. J. *Animal models of human psychology: A critique of science, ethics, and policy.* Seattle: Hogrefe & Huber, 1997.

Singer, P. *Animal Liberation*. New York: Avon, 1975.

Singer, P. *Animal Liberation,* rev. ed. New York: New York Review of Books, 1990.

Sperling, S. *Animal Liberators: Research and Morality.* Berkeley: University of California Press, 1988.

Strand, R., and P. Strand. *The Hijacking of the Humane Movement*. Wilsonville, Ore.: Doral, 1993.

Sutherland, A., and J. E. Nash. Animal rights as a new environmental cosmology. *Qualit. Sociol.* 17: 171–186, 1994.

Index